India Studies in Business and Economics

The Indian economy is considered to be one of the fastest growing economies of the world with India amongst the most important G-20 economies. Ever since the Indian economy made its presence felt on the global platform, the research community is now even more interested in studying and analyzing what India has to offer. This series aims to bring forth the latest studies and research about India from the areas of economics, business, and management science. The titles featured in this series will present rigorous empirical research, often accompanied by policy recommendations, evoke and evaluate various aspects of the economy and the business and management landscape in India, with a special focus on India's relationship with the world in terms of business and trade.

More information about this series at http://www.springer.com/series/11234

K. Palanisami · Krishna Reddy Kakumanu ·
Udaya Sekhar Nagothu · C. R. Ranganathan

Climate Change and Future Rice Production in India

A Cross Country Study of Major Rice Growing States of India

Springer

K. Palanisami
Emeritus Scientist
International Water Management Institute
New Delhi, India

Udaya Sekhar Nagothu
Norwegian Institute of Bioeconomy Research
Ås, Akershus, Norway

Krishna Reddy Kakumanu
Centre for Natural Resource Management (CNRM)
National Institute of Rural Development & Panchayati Raj (NIRDPR)
Hyderabad, India

C. R. Ranganathan
Department of Computing
Coimbatore Institute of Technology
Coimbatore, India

ISSN 2198-0012 ISSN 2198-0020 (electronic)
India Studies in Business and Economics
ISBN 978-981-13-8362-5 ISBN 978-981-13-8363-2 (eBook)
https://doi.org/10.1007/978-981-13-8363-2

This Springer imprint is published by the registered company Springer Nature Singapore Pte Ltd.
The registered company address is: 152 Beach Road, #21-01/04 Gateway East, Singapore 189721, Singapore

Foreword

Agriculture systems around the world, in general, and in India, in particular, are subjected to severe climatic interannual variability and extreme events such as floods, droughts, floods, and hailstorms, all intensified by climate change. Unless climate change impacts are addressed, our agricultural systems will become more vulnerable and put national food and global food security at risk.

Rice is a water-intensive crop and predominantly grown in irrigated conditions. In India, 96 million tons of rice is produced on 44 million ha. This constitutes about one-fourth of the total food grains produced from all the major cereal crops grown. India still needs to produce more rice to feed an increasing population. Because of the water scarcity, it is important to produce more rice with less water. With the impacts of climate change and variability, this will be a formidable challenge for achieving food security for the country.

A set of new interventions in rice production and management are required to address climate change and its various impacts and to see how they could be effectively developed and upscaled. Several studies have indicated that this is possible but do not address how this can be done holistically at a national level. Moreover, the economics of such adaptation strategies has not been studied in detail. I find that in addition to quantifying the impact of climate change on rice for all the regions, this book also evaluates the economics of management technologies in different rice growing regions. This will be incredibly important for policy analysis and formulation.

This book, a first of its kind, focuses on impacts of climate change on rice production. It displays a region-specific impact of climate change on productivity and production of rice and suggests relevant adaptation strategies to address these impacts. I am sure the key messages from this book will be highly relevant to planning future adaptation strategies not only in India but also in other rice growing regions.

I applaud the authors who have taken so much effort to put together this very valuable information in a readable form. I congratulate the research team for this rich contribution. This book will be of immense help to practitioners and policy makers by suggesting necessary climate-smart adaptation measures not only in India but also in other rice growing regions of the world.

David Molden
Director General
International Centre for Integrated Mountain
Development GPO Box 3226 Kathmandu Nepal
http://www.icimod.org
http://www.facebook.com/icimod
http://www.twitter.com/icimod
http://www.youtube.com/icimod

Preface

Warming of the climate system in recent decades is unequivocal. A warmer climate may lead to an intensification of the hydrological cycle, resulting in higher rates of evaporation, and increase/decrease in precipitation, resulting in with varying intensity and extreme events of rainfall. Monsoon rainfall is considered very important in the Indian subcontinent, and the utility of precipitation primarily depends upon its spatial and temporal distribution. It will increase flood risk during the rainy season and strongly reduce water supplies during the dry season. The consequences of changing weather pattern and hydrological cycle might cause different levels of water scarcity and consequently will have an impact on crop yields by triggering associated factors like pests and diseases. The expected changes in climate will ultimately alter regional agricultural systems, with consequences for food production.

Understanding climate change is a challenge particularly due to difficulties in precisely quantifying and effectively identifying the management options, and this opens a wide range of opportunities for integrating adaptations taking into account the economic and social development aspects. Understanding those challenges and tackling them creatively can make climate change adaptation an important way to help build a more resilient farming system. Therefore, the strategies and tools used for coping with within and between crop season weather variability must be strengthened to adapt to climate change. At the farm level, small-scale farmers are interested in improved soil and land management technologies that can also benefit them with carbon credits. They need technological support to address the challenges in water and soil management.

In recent years, researchers and practitioners are developing quantification methods and identifying appropriate adaptation technologies and strategies to the changing climate. However, these methods were applied in different time periods covering different regions and various crops. Even though technologies/improved practices are available to the farmers, the adoption level is very low which ranges from 2 to 30%. Hence, what kind of policy and institutional support could help address the adaptation is also needed in order to address these issues and challenges relating to climate change and agriculture.

To address these issues and the emerging challenges, a collaborative research study was coordinated by the Norwegian Institute for Agricultural and Environmental Research (Bioforsk) with funding from the Norwegian Embassy, New Delhi. As part of this project, a study was carried out to quantify the impact of climate change on rice productivity and identification of suitable adaptation strategies for 13 major rice growing states of India applying uniform methodology and data set. The study results could provide answers to most of the questions relating to the quantification of climate change impacts on rice and the financial feasibility of the identified adaptation strategies. As an outcome of this in-depth research study, this book has been published.

Chapter 1 of this book gives an overview of climate change, its challenges to ensure water and food security and the need for alternative adaptation strategies to reduce the variability. Chapter 2 discusses in detail the rice area, production and productivity trends in India. Decadal growth rates were explained for the central, southern, northern, western and eastern regions of India. Chapter 3 provides the climate change projections using the available data and narrows down the projections to the rice growing regions of India. Chapter 4 explains in detail the concepts, tools and approaches that are used to quantify the climate change impacts. Chapters 5–9 provide results of the detailed analysis of the perception of the farmers on climate change and the impact of climate change on rice productivity and production in northern, eastern, western, central and southern regions, respectively. Chapter 10 presents the climate-smart adaptation practices and their financial viability at the farm level, and the last chapter summarizes the key findings and provides recommendations for stabilizing the rice productivity and production in future. We believe that the key messages from this book will have more impact in mainstreaming and upscaling the appropriate adaptation strategies at both community and state levels.

The authors acknowledge with thanks the excellent technical support extended by Dr C. Karthikeyan, Professor of Agrl. Extension, Dr R. Venkatraman, Agrl. Economist and Former Director of Planning and Monitoring, Dr K. R. Karunakaran, Professor of Agrl. Economics, and Dr T. Arivelarasan, Post doc Fellow of Tamil Nadu Agricultural University while completing the final draft of this book.

New Delhi, India — K. Palanisami
Hyderabad, India — Krishna Reddy Kakumanu
Ås, Akershus, Norway — Udaya Sekhar Nagothu
Coimbatore, India — C. R. Ranganathan

Contents

About the Authors

K. Palanisami is an agricultural economist and international water resources expert. He was a Visiting Professor at the University of Minnesota, USA, Principal Researcher at the International Water Management Institute (IWMI), South Asia Office, Hyderabad and Director of the Water Technology centre, Tamilnadu Agricultural University. He is a Fellow of the National Academy of Agricultural Sciences (NAAS), India. He has led several national and international research projects on climate change, water and agriculture and published numerous papers and books. He is currently serving as an agricultural and water expert for the Interstate Water Resource Department of the Govt of Telangana, India.

Krishna Reddy Kakumanu is an Associate Professor at the Centre for Natural Resource Management at the National Institute of Rural Development and Panchayati Raj, Hyderabad, India. He completed his Ph.D. in Agricultural economics from Justus Liebig University, Germany and worked as a Regional Researcher at the International Water Management Institute (IWMI), Hyderabad & New Delhi, India. He has broad expertise in the economics of conjunctive use of water, energy pricing and climate change impact assessment and adaptation in agricultural and water sectors in India. He has 10 years of research experience and has published several research papers in international journals and books.

Udaya Sekhar Nagothu is a Research Professor and Director (Centre for International Development) at the Norwegian Institute of Bioeconomy Research, Norway. He holds an M.Sc. in Natural Resources Management and Ph.D. in Development Studies from the University of Biosciences, Norway. He has more than 25 years of research and development experience in natural resource management and environment related areas. He has coordinated several large interdisciplinary projects on climate change, water resources management and food security in various countries. Sekhar has published articles in several international journals, and edited and contributed to six books on climate change, sustainable agriculture and food security.

C. R. Ranganathan is an Adjunct Professor at the Department of Computing, Coimbatore Institute of Technology, Tamil Nadu. Prior to this, he was working at Tamil Nadu Agricultural University, Coimbatore for 37 years. He has extensive experience in teaching mathematics, statistics and data science and has led several research projects at Tamil Nadu Agricultural University. He has published a number of research articles in national and international journals, authored the book entitled 'A First Course in Mathematical Models of Population Growth' and co-authored the book 'Climate Change and Agriculture in India'.

Chapter 1
Climate Change and Agriculture in India

K. Palanisami, Krishna Reddy Kakumanu, Udaya Sekhar Nagothu and C. R. Ranganathan

1.1 Introduction

Climate change is a continuous phenomenon, and its variability is a complex phenomenon. In order to assess its impact and to develop appropriate adaptation measures, an interdisciplinary approach is essential. A multi-level assessment and high-resolution data analysis are required to map future climate scenarios and their impacts with reduced uncertainty. This includes both natural and social science-related data, including climate and weather, soils, water availability, cropping patterns, productivity and socio-economic variables that are necessary for impact assessment. Recent literature shows that socio-economic factors are highly relevant as they determine the magnitude of impact on the agriculture sector in the event of extreme weather or climatic event. During the twentieth century, evidences through various studies showed an increase in global temperature and changes in rainfall patterns and rates (IPCC 2001; IPCC 2018; Jung et al. 2002). The IPCC (2007) has projected a rise in temperature of 0.5–1.2 °C by 2020, 0.88–3.16 °C by 2050 and 1.56–5.44 °C by 2080 for South Asia. Similarly, changes in precipitation could be observed as the average precipitation is expected to increase globally (IPCC 2001). However, the magnitude of changes in precipitation among various regions varies with models, i.e. from 0 to 50%, where the direction of change is very strongly indicated and between −30 and +30% where it is not.

Further, the IPCC, in its recent report (2018) indicated that the average global temperature would rise by 1.5–2.0 °C by 2100. As a consequence, the climate-related risks to human security, health, livelihood, food security, water supply and economic growth are projected to increase with global warming.

In India, based on a multi-model climate change, it is likely the country will be more warmer in the range of 1.7–2 °C by 2030 s and 3.3–4.8 °C by 2080s. In the same trend, compared with baseline figure (1961–1990), the precipitation would also tend to increase from 4 to 5% by 2030s and subsequently from 6 to 14% by 2080s (Chaturvedi et al. 2012). In their study, Bal et al. (2016) show projections of maximum temperature and rainfall from six models running over a period of 1970–2100. The

K. Palanisami et al., *Climate Change and Future Rice Production in India*, India Studies in Business and Economics,
https://doi.org/10.1007/978-981-13-8363-2_1

Table 1.1 Projected temperature and precipitation changes in India: 2070–2099

Region	Jan–March	April–June	July–Sept	Oct–Dec
Change in temperature (°C)				
North-east	4.95	4.11	2.88	4.05
North-west	4.53	4.25	2.96	4.16
South-east	4.16	3.21	2.53	3.29
South-west	3.74	3.07	2.52	3.04
Change in precipitation (%)				
North-east	−9.30	20.30	21.0	7.5
North-west	7.20	7.10	27.2	57.0
South-east	−32.9	29.70	10.9	0.7
South-west	22.30	32.32	8.80	8.5

Source Cline, 2007

study concludes that a rise in temperature would be between 2.5 and 4.4 °C and rainfall would increase in the range between 15 and 24% by the end of the century. Similarly, the IPCC (2014) had projected that in India during the twenty-first century, the rainfall will increase by 10–12% with more frequent and heavy rainfall days while the mean annual temperature will rise by 3–6 °C (Table 1.1).

The water is one of the primary resources that is affected by the climate change. India is predominantly arid and semi-arid regions. Hence, any shortfall in water supply, due to climate change, will enhance competition for water use among economic, social and environmental applications. Besides, population growth with improved living conditions would also increase the demand for food, thereby increasing the demand for water many folds, particularly, in the river basins (UNESCO-WWAP 2009). Among the various utility levels, a major portion of available water is used for agriculture. Such demands pose a big challenge to water management. The situation will become more critical, at the time of unfavourable weather or climatic conditions. The climate change in other regions is also expected to impact negatively and by 2050, more than half of all food crops in sub-Saharan Africa and at least 22% of the world area under important food crop, i.e. rice, will be affected (Campbell et al. 2011).

1.2 Climate Change Impacts on Water and Agriculture

Agriculture system in India is subject to severe climatic inter-annual variability, viz. droughts, floods, storms, etc. The system will become more vulnerable under different scenarios of climate change. Due to climate change, among various crops, rice, a major cereal crop would be highly affected. Around 95% of the cultivated area under rice is grown in irrigated conditions and the crop, depending on the soil

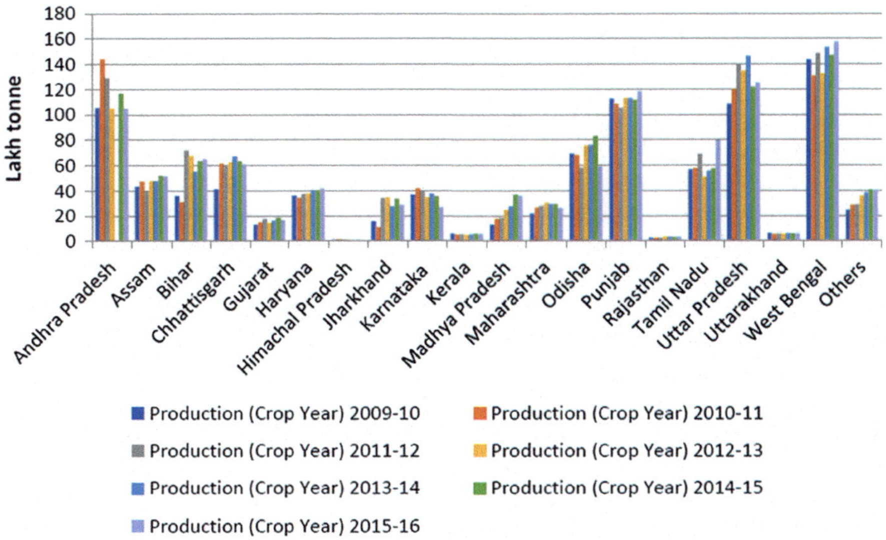

Fig. 1.1 State-wise production of rice (2009–2016) *Source* Indiastat 2013, and agricultural statistics 2014, 2016. *Note* 10 lakhs = 1 million

texture, structure and profile conditions, demands about 1200–2500 mm of water (Reddy and Reddi 1995). In India, rice, with a production contribution of 96 million tons from 44 million ha, constitutes 24% of the total food grains produced from all the major cereal crops grown (Fig. 1.1). To meet the growing food demand, India needs to produce at least 130 million tons of rice by 2030 (Gujja and Thiyagarajan 2009). The projected climate scenario for mid (2021–2050)—and end centuries (2071–2100) in Indian river basins also indicates a change in the future water availability and climatic factors. Under these situations, farmers need to adapt new technologies to meet the demand for rice (Palanisami et al. 2011).

The compounded climate factors would lead to a decline in productivity of the crop. Consequently, an increase in the price of many of the important agricultural crops could take place. Studies show that a rise in temperature during the last few years has caused an overall loss in the crop production in the country. Some studies also indicated with an increase in temperature by 2080–2100, and the probability of loss in crop production in India would be 10–40% (Aggarwal et al. 2008). Using crop simulation models (CERES-Rice and ORYZA1 N), Aggarwal and Mall (2002) and Aggarwal et al. (1997) showed that an increase from 1 to 2 °C temperature without any increase in CO_2 would result in 3–17% decline in rice grain yield in different regions.

Further, under climate change conditions, Mall et al. (2006), based on crop simulation modelling in India, evidenced a decline in the yields of important cereal crops like rice and wheat. Barnwal and Kotani (2010) also found a negative effect on rice yield and yield variability due to climate change in India. Pathak et al. (2003)

reported that during the growing season, the grain yield of rice would decline by 10% for every 1 °C increase in the temperature above 32 °C. Aggarwal et al. (2009) also reported a decline in grain yield in rice to the tune of 5.4, 7.4 and 25.1% with increases in temperatures by 1, 2 and 3 °C, respectively.

Palanisami et al. (2009) also projected the effect of climate variables on three major crops, namely paddy, sugarcane and groundnut in Tamil Nadu. The results showed a reduction in both area and yield of major crops by about 5.2–9.5% due to the impact of climate change. Consequently, overall production would decrease in the range from 9.00–22.00% for these crops in 2020. The National Action Plan on Climate Change (GOI, 2004) identified that climate change will contribute to significant reductions in crop yields and decline in the area cropped in arid and semi-arid zones of India.

This implies a direct impact on the food security in India where rice is the staple diet. Despite the constraints of water scarcity and to meet the demand for rice due to population growth and rapid economic development, it is important to sustain rice production by increasing rice productivity. Producing more rice with less water is therefore a formidable challenge for achieving food security for the country.

So far, the popular method in rice cultivation was transplantation using ponded water that ensured steady yields (Chen et al. 2009). With growing climate uncertainty, insufficient irrigation water both in area commanded by canal and groundwater increased labour requirement and increased emissions, and such practices may not be possible in the future. The rise in temperature could negatively impact the rice yields as they are grown to their threshold (Kelkar and Bhadwal 2007). Further, the requirement of labour is also comparatively higher in transplanted rice cultivation and acts as a constraint. It is estimated that in rice cultivation, about 25–37 labourers (mandays) per hectare are required for transplanting, about 50 mandays per hectare for weeding and about 25 mandays per hectare for harvesting (Technical program 2011; Prakash et al. 2013). The availability of labour is also a major limiting factor in most of the Indian states, as rural people migrate to nearby cities and towns for higher wages or being hired in the government social welfare programmes (e.g. the National Rural Employment Guarantee programme). Hence, the labour cost has significantly increased during recent years (Farooq et al. 2011) and as a consequence, farm mechanization is catching up fast in many of the rice growing regions in India.

Another major concern among scientists working on climate change is the methane emissions from rice fields that contribute to global warming (Pathak et al. 2013). The emissions from flooded fields are found to be higher than those from dry land (Komiya et al. 2010). The methane is emitted when organic matter decays in anaerobic conditions. Hou et al. (2000) showed that the factors affecting methane and nitrous oxide emissions are soil temperature and soil redox potential, net irradiance and organic matter content. The fourth assessment report of IPCC (2007) also indicated that 50% of the methane is emitted from agriculture and 10–20% of these emissions come from rice production alone (Reiner and Aulakh 2000). Overall, the South and East Asian countries are responsible for 82% of methane emissions from rice production. From India alone, annually 4.5 million tonnes of methane is being emitted from rice cultivation (PepsiCo International 2009). There is a need to shift towards more climate-resilient rice cropping systems that can help in methane reduction (Lakshmanan et al. 2009).

Hence, considering the above challenges and to sustain rice production vis-à-vis ensuring food security, new adaptation strategies have to be developed and up-scaled.

Several studies had indicated that delay in onset of monsoon, insufficient canal irrigation water at the tail ends of the command area, additional labour requirement and costs are driving the farmers to adapt to different management practices like Direct Seeded Rice (DSR), System of Rice Intensification (SRI), modified SRI, machine transplanting, change in crop varieties, date of sowing/planting, alternate wetting and drying (AWD) method of irrigation, supplemental irrigation, integrated nutrient, pest management, etc. (Chapagain et al. 2011: Joshi et al. 2013; Mahajan et al. 2013; Li et al. 2014; Kakumanu et al. 2018). But these practices are not adopted fully and in large scale due to lack of timely and insufficient scientific information, non-availability of inputs and other institutional constraints. Moreover, the economics of such adaptation strategies has not been studied in detail. One of the main focuses of this book will be to evaluate the economics of such technologies in different agro-ecological situations.

1.3 Relevance to Policy

The results from this study will be highly relevant for future adaptation strategy planning not only in the 13 states where the study was conducted, but also in other rice growing regions. As rice is one of the main staple crops grown in the country, governments have to seriously consider planning for the future, provide necessary infrastructure and build capacity of the smallholders growing rice. The policy should be supported by evidence-based research results. This book provides necessary recommendations that will be useful for policymakers at the state and national levels. The different chapters in the book summarize results specific to the rice growing regions for the benefit of policymakers and practitioners. Some of the future strategies indicated by the Central Rice Research Institute (CRRI), India, in its Vision 2050 (CRRI 2013) are addressed in this book thus paving the way for a future line of research and policy interface focusing on climate change and rice. These strategies mainly include: (a) research emphasis on improving water and nutrient use efficiencies with special focus on conservation agriculture, climate resilience, rice and rice-based cropping and farming systems, (b) management of knowledge related to rice, with due attention on extension services and fostering linkage and collaboration among organizations namely public, private, national and international, and (c) capacity building of scientists, farmers and other stakeholders to be able globally competitive and to ensure food and nutritional security of the country.

1.4 Outline of the Book

Studies relating to climate change and impacts on rice are scattered and have not focused much on the economics of the impacts and adoption. Researchers in the past have used different approaches (ranging from micro-level field experimentation to macro-level simulation models) for impact assessment. The book integrates and analyses a large data set from different regions and states growing rice in India, and presents the scenarios in a comprehensive manner. The study is one of the first of its kind in the region focusing on the economics of the impacts and adoption. The book addresses the impact of climate change on rice using a uniform methodology/analytical approach and data set related to 13 major rice growing states of India which are grouped into five rice growing regions. The book displays a region-specific impact of climate change on productivity and production of rice and suggests relevant adaptation strategies to address these impacts.

The book contains 11 chapters. The Introduction chapter gives an overview of climate change, its challenges to ensure water and food security, and the need for alternative adaptation strategies to reduce the variability. The second chapter discusses in detail the rice area, production and productivity trends in India. Decadal growth rates were explained for the northern, central, western, eastern and southern regions of India. The third chapter provides the climate change projections using the available data and narrows down the projections to the rice growing regions of India. Chapter four explains in detail the concepts, tools and approaches that are used to quantify the climate change impacts. The chapters five to nine provide results of the detailed analysis of the perception of the farmers on climate change and the impact of climate change on productivity and production of rice in northern, eastern, western, central and southern regions, respectively. Chapter ten presents the climate smart adaptation practices and their financial viability at the farm level and last chapter summarizes the key findings and provides recommendations for stabilizing the rice productivity and production in the future.

The strength of the book, therefore, lies in its in-depth focus on climate change, variabilities and their impacts on productivity and production of rice and the appropriate adaptation strategies to address them. Analysis of the economics of various adaptation strategies relevant to different rice growing regions has added value to the book. Thus, by addressing the climate change and variability impacts, the book downsizes the climate change scenarios at state and region levels and uses relevant econometric modelling methods to integrate and analyse natural and social science-related data in the process. The book will therefore help practitioners and policy-makers by suggesting necessary climate smart adaptation measures for stabilizing rice yields and production in short term and long term.

Chapter 2
Rice Production in India: Analysis of Trend, Constraints and Technologies

D. Suresh Kumar

2.1 Introduction

Rice is the staple food for more than 50% of the population across the globe. As a labour-intensive crop, the rice cultivation provides livelihood and employment to millions of people. The Green Revolution enabled many countries across the globe to increase the production, and in India besides increasing production, the country entered into a new era of input use. The usage of high-yielding varieties, fertilizers and plant protection chemicals has increased manifold particularly in irrigated agriculture. The long-term analysis on area under rice, however, indicated that area is almost stagnated during the past couple of decades. It is estimated that around 130 million tons of rice is needed to feed the population by 2030. With ever-increasing demand for land and water due to urbanization and industrialization, there is only little scope for expanding the area under rice. In addition, the other major production constraints that pose challenges across states are labour and water. Hence, there is a need to increase productivity (Gujja and Thiyagarajan 2009). Evidences show that the irrigated rice consumes 34–43% of the world's irrigation water (Bouman et al. 2007). It is estimated that the total value of rice production in the world is more than US $150 billion per year (Shetty et al. 2013).

The Green Revolution technologies including adoption of HYVs, fertilizers, pesticides and so on have contributed significantly to achieve increased rice yield in many countries. It is evident that India with a total rice area of 44 million hectares

The chapter was commissioned by Dr. Palanisami to add different angles to this book. The author would like to thank Dr. Palanisami for providing substantial inputs in preparing this chapter, without which, the chapter would not have taken the shape that it has now.

D. Suresh Kumar (✉)
Department of Agricultural Economics, Tamil Nadu Agricultural University, Coimbatore, India
e-mail: rithusuresh@yahoo.com

K. Palanisami et al., *Climate Change and Future Rice Production in India*, India Studies in Business and Economics,
https://doi.org/10.1007/978-981-13-8363-2_2

(26.3% of the world rice area) ranks first in area and ranks second in rice production (169 million tonnes) during 2017.

India is not in a good position in terms of rice productivity. Recent estimates show that India's rice productivity (3590 kg/ha) is lower than many rice-producing countries like China (6686 kg/ha), Bangladesh (4219 kg/ha) and Myanmar (4081 kg/ha). Though India has achieved significant increase in productivity of rice from 1.1 tons/ha in 1965 to 2.33 t/ha in 2011, still the productivity is much lower than countries like Egypt, China, Japan and Korea whose rice productivity ranges from 3.70 to 9.00 tons/ha. About 90% of the rice produced in India is used for domestic consumption. Being a carbohydrate-rich cereal, it fulfils 30% of the calories in dietary requirements (Mclean et al. 2002). Burgeoning population is the major driver for ever-increasing demand for rice. It is estimated that the demand for rice will be 113.3 million tons by the year 2021–22 (Kumar et al. 2009). In order to meet the growing demand for rice both for domestic consumption and export demand, it is essential that the rice productivity and production should be increased to a higher level.

Keeping the above issues in view, this chapter attempts to examine the growth rate (compound growth rate) and trend in rice production and productivity across Indian states. Such analysis will help in targeting appropriate policies to boost rice production and productivity in poorly performing states.

2.2 Trend in Area, Production and Productivity

2.2.1 Trend in Area Under Rice

With wide variations in agro-economic conditions and socio-economic diversity, the issue on regional differences in production and productivity of rice is of crucial interest. Therefore, a proper understanding of dimensions of regional differences in area, production and productivity of rice is essential for effective regional planning. The regional differences in area of rice are presented in Table 2.1 (Figs. 2.1, 2.2, 2.3, 2.4 and 2.5).

India has registered only a little growth in area under rice. The rice area has increased from 38.64 million hectares in 1970s to 43.08 million hectares during 2010s. The compound growth rate (CGR) was only 0.369% per annum. There are regional differences in area under rice across regions. Among the southern states, except Tamil Nadu, Andhra Pradesh and Karnataka have registered positive growth in terms of area though the growth rate is lower. In Tamil Nadu, the area under rice has registered a negative growth of 0.975%. This is mainly due to water scarcity, labour scarcity, poor functioning of irrigation tanks and reduction in tank irrigated area and so on.

Among the states in central India, except Madhya Pradesh, Maharashtra state has witnessed a slight increase in area and positive growth rate. There is a small negative

Table 2.1 Decadal trend in rice area across states in India (million ha)

Region/states	1970s	1980s	1990s	2000s	2010s	Over all period 1970–71 to 2012–13
Northern region						
Punjab	0.670	1.582	2.232	2.630	2.833	1.852
	(12.456)	(5.392)	(2.491)	(0.889)		(4.403)
Haryana	0.339	0.545	0.830	1.062	1.235	0.731
	(6.636)	(2.396)	(6.147)	(2.008)		(3.835)
Uttar Pradesh	4.700	5.332	5.602	5.691	5.822	5.365
	(1.339)	(0.027)	(0.822)	(−0.441)		(0.577)
Central region						
Madhya Pradesh	4.611	4.945	4.925	1.637	1.714	3.867
	(0.877)	(0.358)	(7.386)	(−1.329)		(−2.938)
Maharashtra	1.402	1.505	1.524	1.522	1.536	1.491
	(1.686)	(−0.092)	(−0.72)	(−0.020)		(0.239)
Eastern region						
Bihar	5.265	5.189	4.992	3.436	3.141	4.610
	(0.508)	(0.252)	(0.137)	(−0.859)		(−1.342)
Odisha	4.476	4.259	4.634	4.437	4.091	4.426
	(−0.910)	(0.405)	(−1.33)	(−0.016)		(−0.056)
West Bengal	5.138	5.299	5.859	5.774	5.278	5.500
	(0.050)	(1.120)	(0.587)	(−0.025)		(0.325)
Chhattisgarh[a]				3.720	3.751	3.733
				(−1.093)		(0.163)
Western region						
Gujarat	0.451	0.505	0.605	0.668	0.776	0.572
	(0.266)	(0.406)	(2.024)	(2.937)		(1.413)
Southern region						
Tamil Nadu	2.646	2.153	2.181	1.848	1.832	2.180
	(0.234)	(−2.018)	(0.210)	(0.210)		(−0.975)
Karnataka	1.165	1.149	1.412	1.381	1.4	1.285
	(0.699)	(0.376)	(3.452)	(1.367)		(0.691)
Andhra Pradesh	3.499	3.727	3.827	3.672	4.080	3.709
	(1.910)	(0.539)	(0.394)	(1.181)		(0.253)
All India	**38.639**	**40.654**	**43.201**	**43.404**	**43.081**	**41.586**
	(0.876)	(0.413)	(0.654)	(−0.017)		(0.369)

Note Figures in parentheses indicate estimated compound growth rates
1970s: 1970–71 to 1979–80; 1980s: 1980–81 to 1989–90; 1990s: 1990–91 to 1999–2000
2000s: 2000–01 to 2009–10; 2010s: 2010–11 to 2012–13
[a]Data not available for initial years as state was not formed
Compound growth rates for 2010s not worked out due to limited data

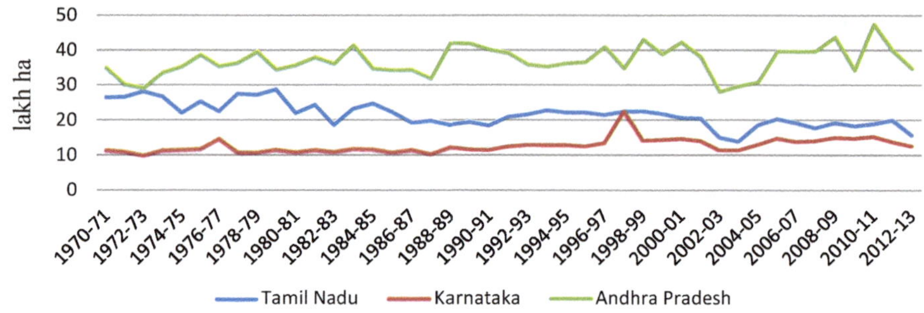

Fig. 2.1 Trend in area of rice in Southern India

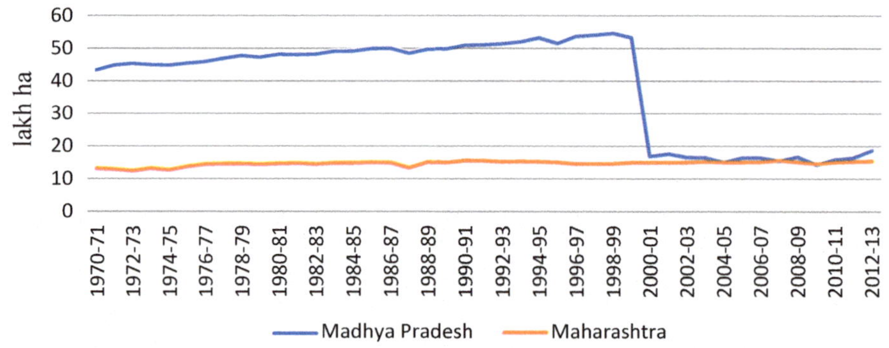

Fig. 2.2 Trend in area of rice in central India

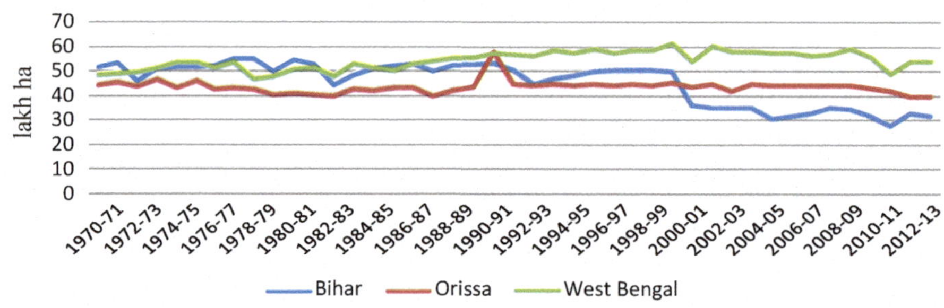

Fig. 2.3 Trend in area of rice in eastern India

growth in Madhya Pradesh over the period of four decades. It is evident that the area under rice in Madhya Pradesh has declined from 4.61 million hectares during 1970s to 1.714 million hectares during 2010s. It is mainly because of bifurcation of Chhattisgarh during 2000 from Madhya Pradesh. As far as eastern India is concerned, slight increase in area under rice is visualized in West Bengal, whereas the other states like Bihar and Odisha have witnessed a decline in rice area.

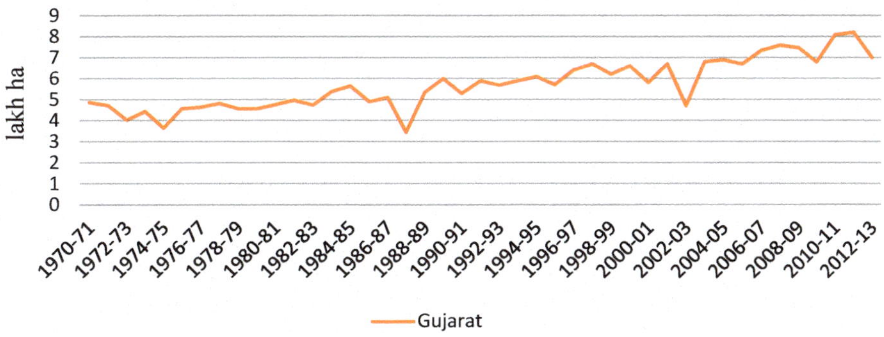

Fig. 2.4 Trend in area of rice in western India

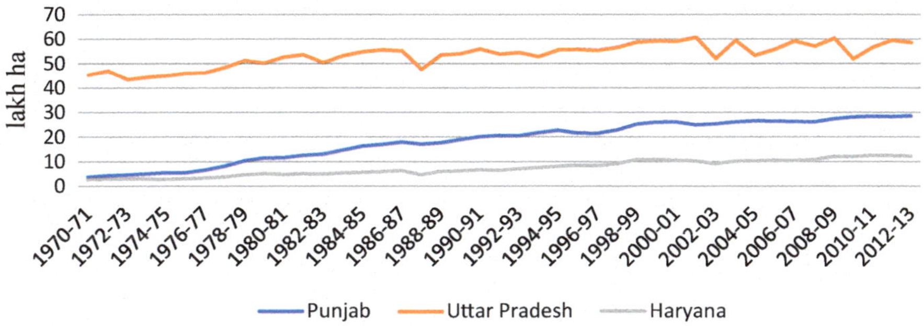

Fig. 2.5 Trend in area of rice in northern India

It is apparent that all states in northern India have witnessed a high growth rate in area during the period from 1970–71 to 2012–13. The states like Punjab and Haryana have registered a positive compound growth rate in rice area. Significant increase in area under rice is seen in Punjab and Haryana. For instance, rice area in Punjab has increased from 0.67 million hectares during 1970s to 2.833 million hectares during 2010s.

2.2.2 Trend in Rice Production

Rice production in India has witnessed a significant progress over the years. Total rice production in the country has increased from 44.76 million tonnes during 1970s to 100.13 million tonnes during 2010s. More than doubling of rice production has been achieved mainly because of Green Revolution and wider adoption of technologies like modern rice varieties, HYV seeds, fertilizers and plant protection chemicals. At all India level, the growth rate in rice production was 2.31% per annum (Table 2.2).

Table 2.2 Decadal trend in rice production across states in India (million tonnes)

	1970s	1980s	1990s	2000s	2010s	Over all period 1970–71 to 2012–13
Northern region						
Punjab	1.675	4.925	7.430	10.000	9.539	6.253
	(18.500)	(6.740)	(2.502)	(2.670)		(5.759)
Haryana	0.697	1.383	2.168	3.083	3.421	1.943
	(11.280)	(2.249)	(4.429)	(4.021)		(4.931)
Uttar Pradesh	4.044	7.211	10.860	11.464	13.311	8.737
	(1.644)	(5.674)	(3.057)	(0.086)		(3.408)
Central region						
Madhya Pradesh	3.302	4.292	5.287	1.485	2.777	3.534
	(−3.064)	(2.032)	(4.956)	(1.573)		(−1.739)
Maharashtra	1.741	2.176	2.322	2.416	3.078	2.227
	(6.932)	(−45.486)	(0.499)	(1.466)		(1.318)
Eastern region						
Bihar	4.678	5.273	6.318	4.569	6.011	5.265
	(0.296)	(4.135)	(4.938)	(−2.278)		(0.222)
Odisha	3.890	4.555	6.287	6.327	7.866	5.446
	(−1.435)	(3.978)	(−4.890)	(4.680)		(1.748)
West Bengal	6.358	8.149	11.797	14.497	13.475	10.446
	(0.779)	(6.828)	(−0.015)	(0.785)		(2.605)
Chhattisgarh[a]				4.742	7.116	5.759
				(−12.968)		(7.663)
Western region						
Gujarat	0.469	0.624	0.903	1.133	1.618	0.840
	(4.140)	(−0.241)	(2.926)	(10.179)		(3.414)
Southern region						
Tamil Nadu	5.182	5.106	6.910	5.353	4.924	5.587
	(0.477)	(3.863)	(1.899)	(−0.838)		(0.227)
Karnataka	2.048	2.238	3.507	3.594	3.546	2.895
	(1.439)	(0.236)	(5.954)	(2.696)		(1.923)
Andhra Pradesh	5.577	8.012	9.726	10.414	11.546	8.649
	(4.202)	(2.514)	(1.640)	(5.675)		(1.843)
All India	**44.759**	**59.775**	**80.338**	**89.166**	**100.127**	**70.715**
	(1.901)	(3.616)	(2.031)	(1.591)		(2.312)

Note Figures in parentheses indicate estimated compound growth rates
1970s: 1970–71 to 1979–80; 1980s: 1980–81 to 1989–90; 1990s: 1990–91 to 1999–2000
2000s: 2000–01 to 2009–10; 2010s: 2010–11 to 2012–13
[a]Data not available for initial years as state was not formed
Compound growth rates for 2010s not worked out due to limited data

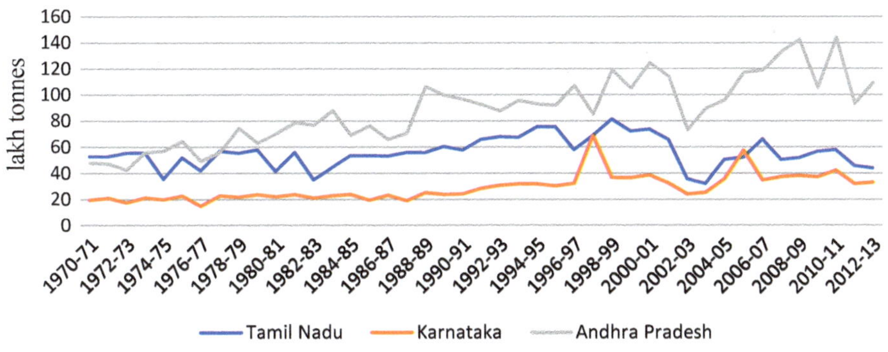

Fig. 2.6 Trend in production in southern India

It is evidenced that all the regions of rice growing areas in the country have distinct behavioural pattern and production conditions, which had implications on overall production. It could be observed that there has been a tremendous increase in rice production across regions of India. In Southern India, a slight decline in production (0.429 million tonnes) was observed in Tamil Nadu during 2010s as compared with the earlier period. There was a decline of 1.557 million tonnes during 2000s when compared to 1990s. The other states, viz. Karnataka and Andhra Pradesh, have registered a positive growth in rice production with a significant increase in production of rice.

In central India, Maharashtra has registered a significant and positive growth in rice production. However, Madhya Pradesh showed a decline in growth rate of 1.74% mainly due to bifurcation of Chhattisgarh from Madhya Pradesh. Similarly, the rice production has witnessed a significant increase in the states of eastern and northern India. Among the states of eastern India, Odisha witnessed a steady growth in production from 3.890 million tonnes during 1970s to 7.866 million tonnes during 2010s. But the most interesting is the unprecedented growth of 13.47 million tonnes recorded by West Bengal during 2010s. The compound annual growth rate of different states rapidly increased during this period (Figs. 2.6, 2.7, 2.8, 2.9 and 2.10).

2.2.3 *Trend in Rice Productivity*

Doubling of rice productivity has been achieved in the country over a period of four decades of development owing to Green Revolution technologies. The productivity level has increased from 1156 kgs/ha during 1970s to 2357 kgs/ha during 2010s (Table 2.3). Though there is a consistent increase in productivity not only in the country but also in all the states, there are significant variations across regions and states. The average rice productivity of states like Punjab (3199 kgs/ha), Tamil Nadu (2627 kgs/ha), Haryana (2556 kgs/ha), Andhra Pradesh (2376 kgs/ha), Karnataka

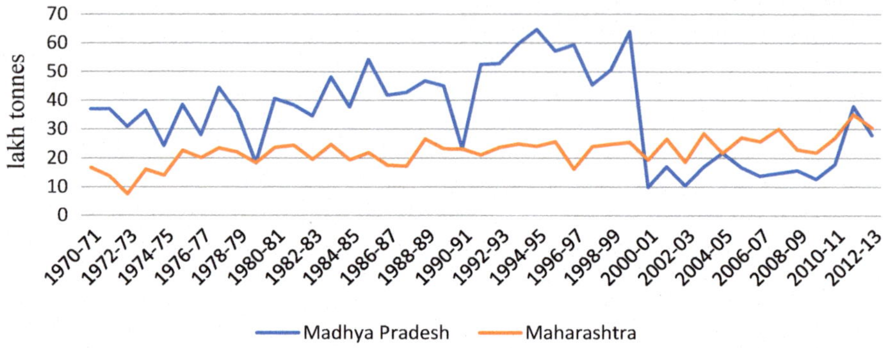

Fig. 2.7 Trend in production of rice in central India

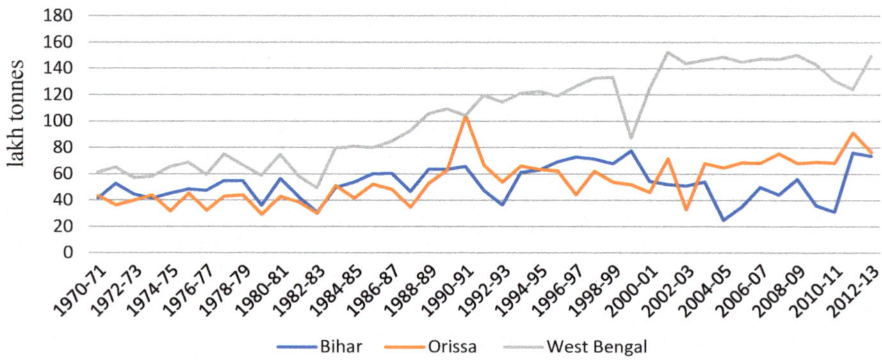

Fig. 2.8 Trend in production of rice in eastern India

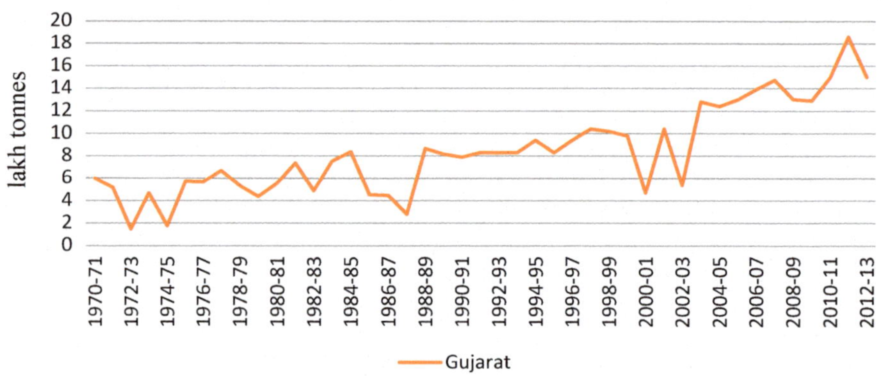

Fig. 2.9 Trend in production of rice in western India

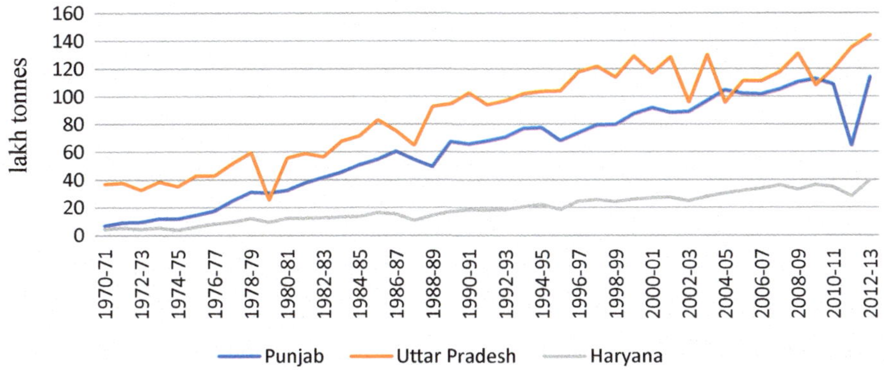

Fig. 2.10 Trend in production of rice in northern India

(2236 kgs/ha) and West Bengal (1929 kgs/ha) had shown productivity levels higher than the country's average level of 1648 kgs/ha. Among the states, the Punjab stands first in terms of productivity with 3199 kgs/ha followed by Tamil Nadu (2627 kgs/ha) and Haryana (2556 kgs/ha).

India registered annual CGR of 1.95% in yield of rice for the period from 1970–71 to 2012–13. Though Indian agriculture has recorded a positive growth rate in rice, there are considerable variations in pattern of growth across regions. The state-wise analysis shows that except Karnataka, all other states have registered positive annual CGR over a period of four decades. Thus, one can speculate that though the area under rice is declining over years, the positive growth in yield in most of the states compensated the reduction in total rice production. The dominating yield effect could be attributed to high level of input use, modern varieties, technologies and agro-climatic factors. For instance, the increase in compounded annual growth rate of rice in Punjab is mainly because of wider adoption of technologies such as semi-dwarf high-yielding varieties and crop production technologies. The low productivity performance at all India level indicates that there is a need to increase the average yield to meet the ever-increasing demand.

The rice productivity in Tamil Nadu has registered a positive growth of 1.34% per annum. The positive growth in yield is a major contributing factor for at least maintaining the rice production in the state, around 5.5 million tonnes, over decades in spite of a negative growth in area under rice in the state (Figs. 2.11, 2.12, 2.13, 2.14, 2.15).

2.3 Constraints in Rice Production

The rice production is subject to many constraints, and these limitations vary across regions. The constraints are technical and socio-economic. The technical constraints

Table 2.3 Decadal trend in productivity of rice across states in India (kg/ha)

Region/states	1970s	1980s	1990s	2000s	2010s	1970–71 to 2012–13
Northern region						
Punjab	2385.600	3092.500	3325.900	3797.600	3852.667	3199.395
	(5.369)	(1.278)	(0.016)	(1.763)		(1.450)
Haryana	2007.600	2534.900	2640.900	2900.700	3031.667	2556.651
	(4.354)	(−0.150)	(−1.641)	(1.979)		(1.145)
Uttar Pradesh	858.400	1349.900	1934.800	2010.600	2312.333	1592.419
	(0.299)	(5.646)	(2.215)	(0.529)		(2.826)
Central region						
Madhya Pradesh	717.400	867.200	1101.400	913.000	1230.000	922.790
	−(3.896)	(1.671)	(−2.271)	(2.907)		(1.015)
Maharashtra	1229.000	1443.400	1591.600	1586.800	1852.667	1489.907
	(5.153)	(−0.579)	(2.115)	(1.494)		(1.046)
Eastern region						
Bihar	886.900	1010.800	1262.500	1319.700	1837.000	1170.000
	(−0.214)	(3.875)	(4.788)	(−1.435)		(1.569)
Odisha	867.400	1064.500	1343.800	1422.600	1653.333	1207.977
	(−0.523)	(3.560)	(3.725)	(4.703)		(1.666)
West Bengal	1238.000	1527.900	2209.600	2509.700	2704.333	1929.419
	(0.829)	(5.642)	(4.203)	(0.803)		(2.435)
Chhattisgarh[a]				1374	1669.667	1443.571
				(−11.991)		(4.801)
Western region						
Gujarat	1017.800	1212.400	1438.478	1665.100	2052.333	1383.599
	(3.873)	(−0.647)	(0.587)	(6.997)		(1.926)
Southern region						
Tamil Nadu	1950.000	2396.500	3164.000	2864.600	3082.667	2627.884
	(0.246)	(6.003)	(0.636)	(−0.303)		(1.338)
Karnataka	1834.300	1945.100	2437.600	2580.200	2733.000	2236.535
	(1.526)	(−0.136)	(2.447)	(−45.162)		(1.199)
Andhra Pradesh	1587.700	2140.500	2539.700	3020.500	3102.333	2376.535
	(2.251)	(1.962)	(1.224)	(1.321)		(1.988)
All India	**1156.300**	**1467.100**	**1858.200**	**2052.300**	**2357.667**	**1648.00**
	(1.013)	(1.371)	(1.371)	(1.606)		(1.948)

Note Figures in parentheses indicate estimated compound growth rates
1970s: 1970–71 to 1979–80; 1980s: 1980–81 to 1989–90; 1990s: 1990–91 to 1999–2000
2000s: 2000–01 to 2009–10; 2010s: 2010–11 to 2012–13
[a]Data not available for initial years as state was not formed
Compound growth rates for 2010s not worked out due to limited data

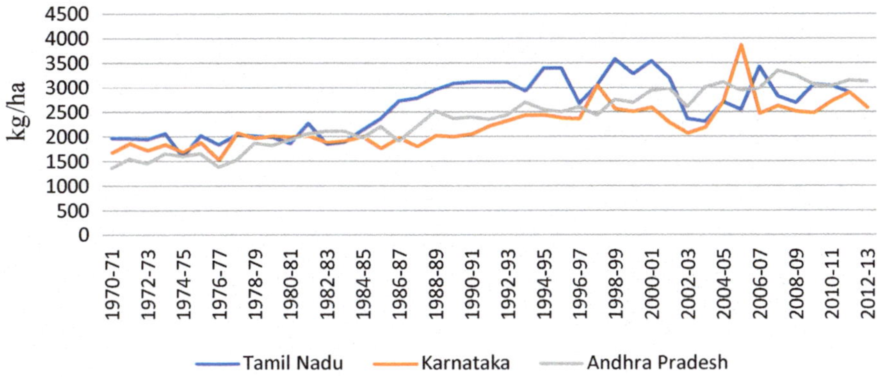

Fig. 2.11 Trend in yield of rice in southern India

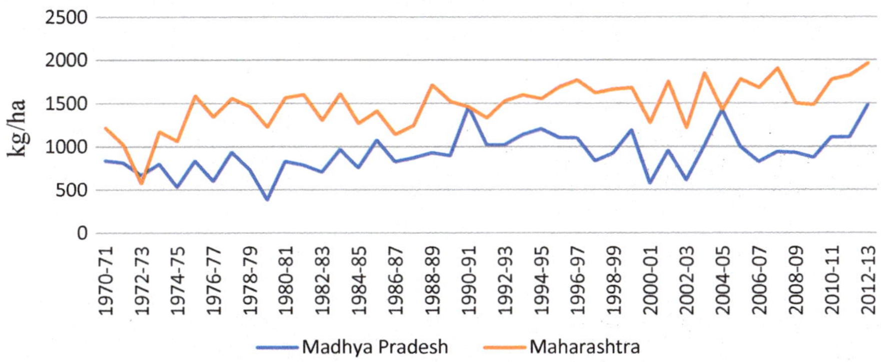

Fig. 2.12 Trend in yield of rice in central India

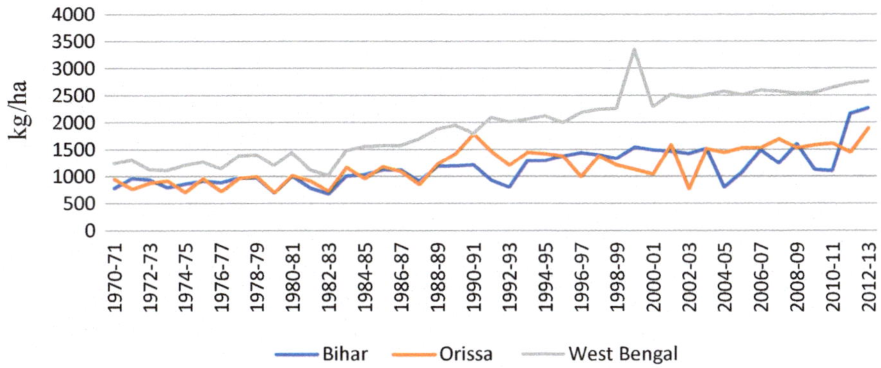

Fig. 2.13 Trend in yield of rice in eastern India

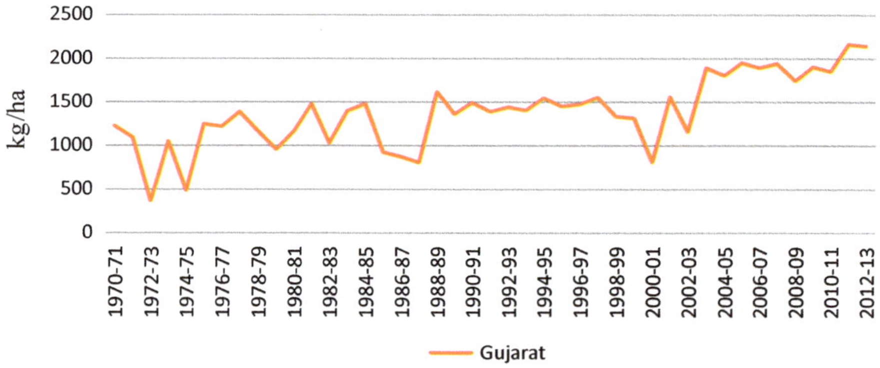

Fig. 2.14 Trend in yield of rice in western India

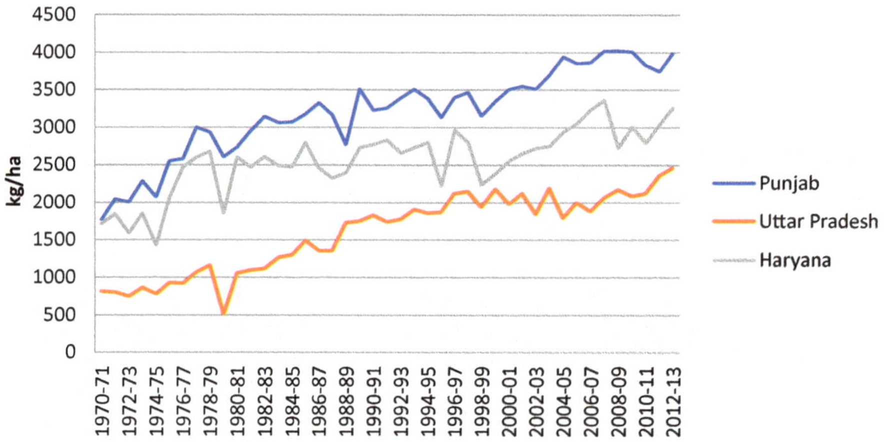

Fig. 2.15 Trend in yield of rice in northern India

include pest and diseases, soil, agronomic, physiological factors, etc. The socio-economic constraints include small size of holding, inadequate supply of inputs, non-availability of inputs, etc.

Studies on rice production across the states however revealed that the production is constrained by many factors like socio-economic, biotic and abiotic factors. The details of different constraints present in different regions are presented in Table 2.4. It is evident that most of the constraints which limit rice production are location-specific and vary across regions.

Andhra Pradesh: The major production constraints in the state are broadly classified into biotic and abiotic constraints (Cheralu, http://www.rkmp.co.in).

Biotic stresses: The insect pests attacking the rice crop are brown plant hopper (BPH), gall midge, stem borer, cutworm, leaf folder, Gundhi bug and rice hispa. The pests like panicle mite and thrips are newly emerging pests of significance in rice.

Table 2.4 Major rice production constraints across states in India

Constraints	States/agro-ecologies
Smallholder-dominant agriculture with poor resource base, suboptimal level of input use	Mostly eastern states of India
Erratic rainfall with poor soils	Madhya Pradesh, Odisha and some parts of Uttar Pradesh
Floods, water logging due to poor drainage	West Bengal, north Bihar and eastern Uttar Pradesh
Poor adoption of modern and high-yielding rice varieties	Mostly eastern states
Suboptimal level of fertilizer use, imbalanced use of nutrients	North-eastern and eastern states
Late sowing and lengthy transplanting period due to monsoon failure and water scarcity	Mostly rainfed lowlands
Low and poor adoption of crop production technologies	Mostly in uplands and lowlands
Saline and alkali soils	West Bengal, Odisha, Andhra Pradesh, Tamil Nadu, Kerala, Karnataka, Maharashtra, Gujarat, Western Uttar Pradesh, Punjab, Haryana, etc.

Source http://www.rkmp.co.in

BPH is a dominant pest of rice specifically in coastal Andhra Pradesh. Recently, it has become widespread throughout the state. In addition to BPH, white backed plant hopper (WBPH) incidence also observed and they occur in mixed population and causing severe damage (hopper burn) to the rice crop. It is estimated that the yield loss due to the WBPH reported to be 10 to 100%. Recently, panicle mite has become an important constraint. Losses caused by the weeds are reduction in yield, loss of nutrients from the soil, poor quality of produce, increased incidence of pests and diseases, competition for resources and increased cost of cultivation. Estimated yield losses due to weeds in low lands are 10–30%, while it is 40–60% in case of rainfed rice.

The abiotic factors that limit rice yields are severe drought, water logging, temperature, zinc deficiency, salinity and so on. In addition, the socio-economic constraints like illiteracy, less inclination to know the latest technologies, labour scarcity, poor resource base and investment form other important constraints.

Karnataka: Poor adoption of modern varieties, poor water management, use of imbalanced nutrition, sudden outbreak of pest and diseases in irrigated areas and moisture stress and poor nutrient uptake in rainfed areas are the major constraints that limit rice production in the state (Rajanna, http://www.rkmp.co.in).

Tamil Nadu: Rice is predominantly grown under irrigated conditions. In some pockets, it is also grown under rainfed conditions. The rainfed rice (upland rice) often subjected to severe drought or high rainfall. In addition, soils in upland rice cultivating areas may not be suited well to rice cultivation and hence led to poor fertilizer–yield response (Thiyagarajan and Kalaiyarasi, http://www.rkmp.co.in).

It is found that the major rice production constraints in the state are: (i) non-availability of HYV seeds which led to cultivation of traditional rice varieties, (upland and rainfed lowland), (ii) poor fertility status of soils due to wind and water erosion, (iii) high incidence of pest and diseases, weeds, etc., (iv) poor germination percentage, poor population and uneven growth, (v) monsoon failure and consequent delayed sowing and lengthy transplanting period.

Shanmugam et al. (2007) found that the top ten rice yield constraints which limit rice production in the state are leaf folder, yellow stem borer, drought, blast, ear head bug, zinc deficiency, RTV, sheath rot, BPH and bacterial leaf blight. The yield loss due to all these constraints are estimated to be 11.40, 9.35, 7.84, 7.14, 7.12, 5.26, 5.20, 4.63, 4.31 and 4.01%, respectively, which account for a total production loss of 1.7 million tonnes.

Madhya Pradesh: The rice production in Madhya Pradesh is limited by various biotic and abiotic constraints. The important diseases include leaf blast, brown spot, bacterial blight, sheath blight, false smut and seed discoloration, and the pests are leaf folder, case worm, Gundhi bug, WBPH, BPH, Stem borer etc. The climatic factors such as poor and erratic rainfall and high temperature are the major constraints. The critical socio-economic constraints include suboptimal and imbalanced use of fertilizers, interrupted power supply to agriculture and inadequate supply of fertilizers. Insufficient research to address the major rice production constraints and poor linkages with the development departments hinder rice production (Rao, http://www.rkmp.co.in).

Maharashtra: In Maharashtra, rice is grown mainly in four major regions, namely Konkan, Marathwada, Western Maharashtra and Vidarbha. Among the four regions, the Konkan region has witnessed maximum productivity contributing around 43% of rice production in the state. Though Vidarbha region has larger extent of area under rice in the state, its productivity is very low when compared to other regions (Thaware et al. http://www.rkmp.co.in).

Rice production in the state is constrained by many diseases and pests. Rice blast, bacterial leaf blight, sheath blight and false smut are the major diseases causing yield loss in rice. Similarly, the major pests are yellow stem borer, BPH, leaf folder, army worm and gall midge. The other important constraints are salinity, poor adoption of HYVs, low soil fertility, inefficient water management, suboptimal level of fertilizer use, dry spells at critical stages and poor resource base of farmers. The eastern part of Vidarbha suffers heavily due to erratic rainfall particularly during kharif season.

Uttar Pradesh: The rice growing seasons in the state are classified as wet season (kharif), winter season (boro) and summer season (zaid). Kharif is the main rice growing season in the state, and more than 98% rice is cultivated during this season covering short-, medium- and long-duration varieties.

The average productivity of rice in the state is low (1592 kgs/ha) which is even lower than the national average. This might be due to the fact that major rice area is under rainfed. The frequent occurrence of various biotic and abiotic constraints is also responsible for the low productivity. Rice blast, bacterial leaf blight, brown spot, sheath blight and neck blast are the major rice diseases that cause significant yield loss. Similarly, pink stem borer and yellow stem borer are the major pests which cause around 40% of yield loss in rice. The other insects including leaf folder, white grub, Gundhi bug and termite also limit the rice yield in the state. In rainfed conditions, weeds and moisture stress are the major constraints limiting rice yields. The nutrient deficiency such as zinc, iron and boron and iron toxicity limit the higher rice yields.

The other factors which constrain higher rice yield in the state are deteriorating soil health conditions due to imbalanced use of fertilizers, inadequate application of organic manures, secondary and micro-nutrients, low seed replacement rate, lack of farm mechanization, inefficient utilization and improper water management, small size of holding, and inefficient extension mechanism for transfer of technologies (Dwivedi, http://www.rkmp.co.in).

Odisha: In Odisha, the biotic constraints such as brown plant hopper, white backed plant hopper, stem borer, gall midge, sheath rot and bacterial leaf blight are found to be most yield reducing constraints. It is also found that lack of HYVs, resistance to stem borer and bacterial leaf blight and tolerance to water logging conditions are the major constraint to higher rice yields in the state (Das http://www.rkmp.co.in).

West Bengal: Rice is an important crop in West Bengal, cultivated in different production environments such as different soil types, water regimes, water logging conditions and so on. In general, there are four major types of production environments in which rice is grown. They include irrigated, rainfed (lowland and upland), coastal saline and flood-prone areas. On an average, the yield gap II, estimated as 1.32 tons/ha, is due to various technical and socio-economic constraints. The major constraints causing the yield gap are: (i) submergence and droughts at early stages of plant growth (ii) important pests like stem borer, brown plant hopper, leaf folder and green leafhopper, (iii) plant diseases such as bacterial leaf blight, brown spot and sheath rot and (iv) soil salinity and weeds as the other constraints. In upland rice-cultivating areas, droughts, stem borer, green leaf hopper, blast, brown spot, acid soils and weeds limit higher rice yields (Adhikari et al. http://www.rkmp.co.in).

Chhattisgarh: The state is not exempted from the constraints that limit rice production. The common constraints prevalent in the state are poor and erratic rainfall, low pH, losses of nutrients particularly nitrogen and potassium due to leaching, poor fertility status, diseases (blast and brown spot), pests like leaf hopper, Gundhi bug, termite, rootknot nematodes and so on. The other constraints which limit rice production in the state are non-availability of high-quality seeds, poor adoption of technologies, etc. (Pandey et al. http://www.rkmp.co.in).

Punjab: Though Punjab state is leading in rice productivity and production among states of India, there is scope for improvement in productivity. The rice production in the state is also affected by different constraints like pests (yellow stem borer, leaf folder, white backed plant hopper, brown plant hopper) and diseases (sheath blight, sheath rot, false smut, foot rot and brown spot) (Rang et al. http://www.rkmp.co.in).

Gujarat: The major diseases which limit to higher rice yields are bacterial leaf blight (BLB), rice blast, grain discoloration and sheath rot. Rice stem borer, brown plant hopper, WBPH, leaf roller and wireworm are the major insect pests in the state.

Weed infestation particularly in upland drilled rice and rice nurseries also are found to limit rice production in the state (Pathak et al. http://www.rkmp.co.in).

2.4 Climate Change and Rice Production

Climate change and its impacts have received much attention in recent decades. It is expected that the climate change may result in significant negative impacts on growth and development, poverty alleviation and achievement of the Millennium Development Goals (MDGs). Consistent efforts taken globally to achieve the developmental objectives may be thwarted by climate change. It is argued by many that the climate change may increase risk, particularly for those who depend on weather patterns, agriculture, water and other natural resources for their livelihoods (El-Ashry 2009). It is found that increase in temperature in India could reduce farm net revenues by 9–25% (http://www.eldis.org/static/DOC11253.htm). Climate variability and change add to the vulnerability in general and those who are already poor in particular. With the climate change on its way, even though mitigation measures are taken up by both Central and State governments, there is a dire need for adapting to the changes like increase in average temperature, aridification of the dry regions and more extreme floods and droughts (Hoff 2004).

Both the Central and State governments have been implementing various adaptive measures. They include policy measures (watershed development programmes, bringing water from the neighbouring states), technological and structural measures (demand-side management of water technologies like drip and sprinkler irrigation, system of rice intensification), change of use activity or location (changes in land use like shifting towards rainfed cultivation, farm diversification such as inclusion of livestock), monitoring and forecasting (weather forecasts, automatic weather forecast station) and risk sharing and spreading (crop insurance, weather-based crop insurance, micro-insurance). In order to prepare for the climate variability, each production environment will require a specific set of adaptation measures addressing various scales of intervention and vulnerable groups.

The climate challenge facing Indian agriculture needs to be addressed in a proper way. It is evident that the annual mean rainfall and also that during monsoon periods decline over the years and the variability in rainfall is also increasing. Another threatening issue is an increase in temperature as well (Table 2.5).

The changes in temperature, radiation, rainfall and CO_2 levels can affect the yield of rice through their direct effect as well as indirect effects such as higher incidence of pests and diseases, and availability of water for irrigation.

A time series analysis of trend in rice yield in the Indo–Gangetic plains shows a declining trend, and this is partly related to the gradual change in weather conditions in the last two decades (Agarwal 2008). The simulation analysis indicated that irrigated rice is likely to lose the yields up to 23% in Upper Ganga Basin. Gupta et al. (2012) observed that climate change is likely to reduce the yields of rice productivity in 16 major agriculture-intensive states of India. It is also found that a 4°

Table 2.5 Details on agro-climatic and weather parameters

1951/52 to 1967/68		1968/69 to 1980/81	1981/82 to 1990/91	1991/92 to 1996/97	1997/98 to 2001/02	2002/03 to 2006/07	2007/08 to 2011/12
Annual rainfall (cm)							
Mean	122.5	118.7	120.1	121.0	118.5	113.7	111.7
Std. deviation	12.5	10.2	11.5	7.2	8.3	9.4	10.0
Coefficient of variation (%)	10.20	8.59	9.58	5.95	7.00	8.27	8.95
Monsoon rainfall (cm)							
Mean	91.9	88.8	88.8	90.0	87.8	83.9	86.6
Std. deviation	10.1	9.6	11.0	6.5	5.5	7.9	9.7
Coefficient of variation (%)	10.99	10.81	12.39	7.22	6.26	9.42	11.20
Annual temperature anomaly from normal (°C)							
Mean	0.04	−0.03	0.09	0.19	0.34	0.56	0.65
Std. deviation	0.28	0.24	0.03	0.10	0.22	0.11	0.26
Coefficient of variation (%)	700.00	−800.00	33.33	52.63	64.71	19.64	40.00

Source Climate bulletins and other publications of the India Meteorological Department Government of India, Twelfth Five Year (2012–2017), Economic Sectors, Volume-II, Government of India, P. 3

rise in temperature will lead to 41% reduction in rice yields (Geethalakshmi et al. 2011). Nandhini et al. (2006) mentioned that rice cultivable land has declined due to scarcity of inputs and scanty rainfall in Tamil Nadu (India). Similarly, Hundaland Kaur (2007) found that 1° increase in minimum temperature leads to a reduction in yield of rice by 3% in Punjab (India). Saseendran et al. (2000) found that change in temperature up to 5 °C can lead to a continuous decline in the yield of rice and every one degree increase in temperature will lead to a 6% decline in yield of rice in Kerala. Haris et al. (2010) found that rice production may decline by 31% in 2080 due to climate change in Bihar.

Researchers found that the grain yield of rice will decline by 10% for every 1 °C increase in temperature above 32 °C (Pathak et al. 2003). The climate change impact on the productivity of rice in Punjab (India) has shown that with all other climatic variables remaining constant, temperature increase of 1, 2 and 3 °C would reduce the grain yield of rice by 5.4, 7.4 and 25.1%, respectively (Aggarwal et al. 2009).

2.5 Summary

This chapter examined the trend in area, production and yield of rice in major rice growing states of India. From the analysis, it is found that there are variations in area, yield and production of rice across regions and states. The overall growth of rice area during the four decades was 0.4%, the growth in rice productivity was 1.95%, and the growth in rice production was 2.36%. Also, there exists significant yield gap in rice production across states and the rice production is constrained by many technical, socio-economic and institutional constraints. In the case of climate change impact, raise in temperature in the crop season has reduced the rice yield.

Significant policy supports will help minimize the yield gap and increase both productivity and production of rice in the long run. Hence, the policy focus may be tilted towards developing and managing the irrigation infrastructure, access to institutional credit, supply of inputs, adoption of technologies particularly use of high-yielding variety seeds, integrated nutrition and pest management, water-saving methods, climate smart rice management practices, development of market infrastructure and adequate price supports.

Appendix

States	Yield gap and its reasons	Constraints	Reasons	Impact of technologies	Ruling varieties/ recommended varieties	Strategies to increase the production of rice crop
Southern						
Tamil Nadu	a. **Maximum Attainable Yield** is the rice yields of experimental/on-farm plots with **no physical, biological and economic constraints** and with the best-known management practices at a given time and ecology b. **Farm-Level Yield** is the average yield of farmers (southern) obtained in a given target area at given time and ecology	Saline, acidic soils and alkali soils Knowledge gaps Use of traditional varieties Poor and imbalanced use of fertilizers Poor adoption of crop production technologies	Difficulty in removal of salts in the coastal region **Toxicity** due to high **pH and** due to the presence of sodium either as carbonate or as bicarbonate in the alkali or saline–alkali soils. Highly dispersed soil under alkaline or saline–alkali situation and drainage is a problem Lack of awareness about high-yielding varieties and hybrid rice technology Continuous use of traditional varieties due to the **non-availability of seeds** and farmers **lack of awareness** about **high-yielding varieties** (upland, rainfed lowland and deep-water areas) **Poor adoption** of improved crop production technology due to **economic backwardness** of the farmers	Encouraging the use of soil ameliorants for improving saline, alkaline and acidic soils Potential of conservation agriculture including zero or minimum tillage, direct-seeded rice, precision agriculture, site-specific nutrient management	IR-64, CO-47, ADT-36, ADT-37, ADT-43, ADT-45, ADT-47, ADT-48, ASD-16, ASD-17, ASD-20	Adoption of SRI cultivation Creating awareness and motivating hybrid rice cultivation through demonstration Improving soil fertility through crop rotation Encouraging balanced use of fertilizers and plant nutrients Adoption of improved water management practices Creating awareness through demonstrations, farmer's field school Attractive Market Price and Insurance coverage

(continued)

(continued)

States	Yield gap and its reasons	Constraints	Reasons	Impact of technologies	Ruling varieties/ recommended varieties	Strategies to increase the production of rice crop
Karnataka	Erratic rainfall Poor in adoption of modern varieties Global climatic changes Poor water management, use of imbalanced nutrition, sudden outbreak of pest and diseases in irrigated areas and moisture stress and poor nutrient uptake in rainfed areas			Integrated nutrient management (organic manures + biofertilizers + chemical fertilizers) Enhance water productivity through wider adoption of SRI Aerobic rice cultivation in water-limited areas like tail-end areas of the canal, tank-fed areas and borewell-irrigated areas if, the cultivation of rice is inevitable **Aerobic rice cultivation** A new mid-early, high-yielding variety MAS 946-1 is recently released by the University for Aerobic Cultivation	MAS 946-1 MAS 26, Rasi, Farmer grown locals Doddabairanelli	Transfer of recent technologies on rice production through proper extension methods like FLDs, Video conferencing, Rice Knowledge Management Portals and extension services supported by Mobile phone based MMSs and SMS Enhance water productivity through wider adoption of SRI, Aerobic and AWD method of cultivation in water-limited areas like tail-end areas of the canal, tank-fed areas and bore Well irrigated areas if, the cultivation of rice is inevitable

(continued)

(continued)

States	Yield gap and its reasons	Constraints	Reasons	Impact of technologies	Ruling varieties/ recommended varieties	Strategies to increase the production of rice crop
Andhra Pradesh	Though farmers adopt HYVs, often they ignore adoption of other crop production technologies and solving other related problems Cyclones and floods are the frequently occurring constraints which limit rice production in the state	Genetic improvement of rice yield and strengthening of quality seed production Promoting hybrid rice cultivation Stabilizing yield through molecular breeding Improving resource use of water Water-saving and yield-enhancing strategies			BPT-5204, Kavya MTU	Collection and maintenance of germplasm bank Identification of suitable areas for hybrid rice seed production in the state Ensuring proper transfer of technology Promotion of vermicompost, green manuring, application of gypsum, zinc sulphate etc. Educating the farmers through capacity building on improved crop management practices likeINM, IPM, Post harvest technology, Water management

(continued)

(continued)

States	Yield gap and its reasons	Constraints	Reasons	Impact of technologies	Ruling varieties/ recommended varieties	Strategies to increase the production of rice crop
Central						
Maharashtra		Reduction in soil organic matter Widening NPK Ratio Micro-nutrient deficiencies Declining yield-fertilizer response in terms of grains/kg fertilizer Heavy metal pollution Low fertilizer use efficiency			Ratnagiri 24, Sahyadri—2, Sahyadri—3	Promotion of Integrated Nutrient Management Adoption of Conservation Agriculture Practices Optimal use fertilizers Evolution of high-yielding, semi-dwarf, fertilizer responsive rice varieties with different duration and grain types suitable for different ecosystem Developing nutrition-rich rice varieties Developing pest and disease resistant varieties

(continued)

(continued)

States	Yield gap and its reasons	Constraints	Reasons	Impact of technologies	Ruling varieties/ recommended varieties	Strategies to increase the production of rice crop
Madhya Pradesh		Poor adoption of hybrid rice Poor adoption of management practices led to only marginal increase in hybrid rice and significant variations across regions Higher seed cost Poor Non-availability of quality seeds in time and at reasonable price			Mahamaya, Ratnagiri 3	Creating awareness through campaigns and demonstration on SRI, aerobic rice etc. Promoting wider adoption of hybrid rice under SRI method and proper water management practices
Eastern						
Bihar	Erratic and uneven distribution of rainfall	Suboptimal level of input use Poor extension services Severe weeds			PA 6444, PHB 71, Pusa RH 10, KRH 2, Rajlaxmi, Sahyadri, DRRH Rajashree, Sita, Rajendra Mahsuri, Prabhat	Promotion of optimal use of different inputs through development programmes and interventions KVKs may be encouraged to ensure proper transfer of technologies and adoption of various crop production technologies

(continued)

(continued)

States	Yield gap and its reasons	Constraints	Reasons	Impact of technologies	Ruling varieties/ recommended varieties	Strategies to increase the production of rice crop
Odisha	Erratic and uneven distribution of rainfall Non-availability of quality seeds and fertilizers in time and at reasonable prices	Poor marketing and credit facilities, poor adoption of modern rice production technologies			Rajalaxmi, KRH-2, PHB-71, PA 6201, PA 6444, PA 6129 and IR 64, IR 36, Pooja, CR 1018, Savitri, Ratna, Khitish, Annada, CR 1017, Lunishree, CR 1014	Soil amelioration by lime (7 lakh ha). Paper mill sludge for acidic soil treatment in Odisha Minikits of saline tolerant varieties More use of quality seeds Promoting farm mechanization wherever possible Developing propoer post-harvest technologies and infrastructure
West Bengal	Yield gap II is not able to manage by farmers	Major rice production constraints are floods and drought in addition to other biotic and abiotic constraints		Development of location-specific rice production technologies Evolving varieties for drought-tolerant/flood-tolerant/salt-tolerant high-yielding rice varieties for flood-prone upland, salinity affected coastal land and flood-prone lowland ecosystems, respectively Adoption of **rain water harvesting structures** Adopting IPNS technique Encouraging organic farming and biofertilizers application Pagel	Khitish, Satabdi, Sabita, Shashi, Bipasa	Developing **post-harvest technologies** Promotion of public-private entrepreneurship Promoting **Seed Village' programme** Adoption of low-cost, resource conserving **eco-friendly technologies**

(continued)

(continued)

States	Yield gap and its reasons	Constraints	Reasons	Impact of technologies	Ruling varieties/ recommended varieties	Strategies to increase the production of rice crop
Chhattisgarh	Poor and imbalanced use of fertilizers	Water stress and flooding, poor drainage Suboptimal use of fertilizers and plant nutrients Socio-economic Technological	Light texture of soil, shallow soil depth, un-bounded Sloppy and undulating topography Accumulation of water in lowland areas creates hurdles in adoption of management practices Poor and uneven distribution of rainfall Resource poor farmers Insecurity of land tenure Inequality in land ownership Labour scarcity, non-availability of inputs in time Poor adoption of improved crop production technologies Non-availability of high-quality seeds in time Lack of suitable rice varieties Lack of suitable machinery		IR 64, IR 36, Mahamaya, Kranti, Purnima, Swarna, MTU 1010, Danteswari, Bamleshwari	Adoption of improved crop production technologies and efficient water management technologies

(continued)

(continued)

States	Yield gap and its reasons	Constraints	Reasons	Impact of technologies	Ruling varieties/ recommended varieties	Strategies to increase the production of rice crop
Western						
Gujarat		Blast, Grain Discoloration and Sheath Rot are the major diseases in the state. Whereas; Stem Borer, Brown Plant Hopper, WBPH, Leaf roller and Wire worm are the major insect pest in the Gujarat Salinity and drought are the major abiotic constraints in rice cultivation		SRI method may be encouraged to ensure higher rice yields	GR-4, GR-5, GR-7, GR-12, NAUR-1, IR-28	Assured **canal water supply** at the time of raising nurseries and during dry spell may be ensured Treatment of seedlings with biofertilizers prior to transplanting **SRI and aerobic method** of cultivation may be encouraged **Water harvesting** in rainfed rice Evolving**suitable hybrid rice** Creating awareness through circulation of print outs on different technologies, conducting demonstrations etc.

(continued)

(continued)

States	Yield gap and its reasons	Constraints	Reasons	Impact of technologies	Ruling varieties/ recommended varieties	Strategies to increase the production of rice crop
Northern						
Punjab	Yield gap is due to various technical and socio-economic constraints	Major constraints for these ecosystems are (i) Submergence and droughts at seedling stage (ii) stem borer, brown plant hopper, leaf folder and green leafhopper as the major insects; (iii) bacterial leaf blight, brown spot and sheath rot as the major diseases and (iv) soil salinity and weeds as the other constraints Labour shortage			Basmati 217, Basmati 370, Basmati 385, Basmati 386, Punjab Basmati 1, Super Basmati	Adoption of improved land, crop water management practices
Haryana	Fluctuating water tables, salinity, declining soil health and low crop productivity				CSR 30 and CSR 27 Pusa Basmati 1, Pusa Sugandh 4, Pusa Sugandh 5 and HBC 19 Taroari Basmati	Adoption of improved crop production technologies
Uttar Pradesh		Problem soils Poor adoption of crop production technologies Erratic rainfall, water logging and floods	Suboptimal of input level of input use, unscientific crop rotation and poor use of organic manures poor adoption of modern technologies	Adoption of INM Effective technology transfer	Narendra Dhan 3112-1, Prakhar, NDR 2064, Narendra Usar Sankar Dhan 3	Encouraging optimal use of fertilizers and inputs Extension mechanism may be strengthened

Source Status paper on Rice for different states, Rice Knowledge Management Portal, Directorate of Rice Research, Rajendranagar, Hyderabad 500030. Field surveys
Visit Rice Knowledge Management Portal http://www.rkmp.co.in

Chapter 3
Climate Change Projections for Major Rice Growing States of India

S. Senthilnathan

This chapter presents evaluation of the model baselines with respect to the observations. Further, it also describes the methodology and approach used for extracting future climate change projections for the major rice growing states to predict the rice production in India.

3.1 Selection of Global Climate Models for Climate Change Projections

A detailed climate projection has been derived from the global coupled model results now available from Coupled Model Intercomparison Project 5th Phase (CMIP5). The 5th IPCC Assessment Report (AR5) was based on analysis of CMIP5. Results of multi-member ensembles of present-day and future climate change scenario simulations from over 25 models run at leading modelling centres throughout the world are now available. Modelling groups across the world have contributed data from more than one model versions. Table 3.1 gives a list of the recent CMIP5 models with daily and monthly data sets available (http://cmip-pcmdi.llnl.gov/cmip5/data_portal.html). In CMIP5, a detailed diagnostics of the historical (current climate) simulations performed by these models indicate that eight of them are capable of simulating the regional distribution of mean monsoon precipitation and its spectrum of variability over South Asia during south-west monsoon season (June through September)

The chapter was commissioned by Dr. Palanisami to add different angles to this book. The author would like to thank Dr. Palanisami for providing substantial inputs in preparing this chapter, without which, the chapter would not have taken the shape that it has now.

S. Senthilnathan (✉)
Department of Agricultural Economics, Tamil Nadu Agricultural University, Coimbatore, India
e-mail: senthilnathanagri@gmail.com

K. Palanisami et al., *Climate Change and Future Rice Production in India*, India Studies in Business and Economics,
https://doi.org/10.1007/978-981-13-8363-2_3

Table 3.1 List of GCMS participated in CMIP5 modelling groups

S. No.	Modelling centre	Institute ID	Model name
1	Commonwealth Scientific and Industrial Research Organization (CSIRO) and Bureau of Meteorology (BOM), Australia	CSIRO-BOM	ACCESS1.0
2	Beijing Climate Center, China Meteorological Administration	BCC	BCC-CSM1.1
3	College of Global Change and Earth System Science, Beijing Normal University	GCESS	BNU-ESM
4	Canadian Centre for Climate Modelling and Analysis	CCCMA	CanESM2 CanCM4 CanAM4
5	National Center for Atmospheric Research	NCAR	CCSM4
6	Community Earth System Model Contributors	NSF-DOE-NCAR	CESM1(CAM5)
7	Centre National de Recherches Meteorologiques/Centre European de Recherche et Formation Avancees en CalculScientifique	CNRM-CERFACS	CNRM-CM5
8	Commonwealth Scientific and Industrial Research Organization in Collaboration with Queensland Climate Change Centre of Excellence	CSIRO-QCCCE	CSIRO-Mk3.6.0
9	EC-EARTH Consortium	EC-EARTH	EC-EARTH
10	LASG, Institute of Atmospheric Physics, Chinese Academy of Sciences and CESS, Tsinghua University	LASG-CESS	FGOALS-g2
11	LASG, Institute of Atmospheric Physics, Chinese Academy of Sciences	LASG-IAP	FGOALS-gl FGOALS-s2
12	The First Institute of Oceanography, SOA, China	FIO	FIO-ESM

(continued)

Table 3.1 (continued)

S. No.	Modelling centre	Institute ID	Model name
13	NOAA Geophysical Fluid Dynamics Laboratory	NOAA GFDL	GFDL-CM3[a] GFDL-ESM2G GFDL-ESM2 M
14	NASA Goddard Institute for Space Studies	NASA GISS	GISS-E2-H GISS-E2-R
15	National Institute of Meteorological Research/Korea Meteorological Administration	NIMR/KMA	HadGEM2-AO
16	Met Office Hadley Centre	MOHC	HadCM3 HadGEM2-CC[a] HadGEM2-ES
17	Institute for Numerical Mathematics	INM	INM-CM4
18	Institute Pierre-Simon Laplace	IPSL	IPSL-CM5A-LR IPSL-CM5A-MR
19	Japan Agency for Marine-Earth Science and Technology, Atmosphere and Ocean Research Institute (The University of Tokyo), and National Institute for Environmental Studies	MIROC	MIROC-ESM MIROC-ESM-CHEM
20	Atmosphere and Ocean Research Institute (The University of Tokyo), National Institute for Environmental Studies, and Japan Agency for Marine-Earth Science and Technology	MIROC	MIROC4 h MIROC5
21	Max Planck Institute for Meteorology	MPI-M	MPI-ESM-LR MPI-ESM-P
22	Meteorological Research Institute	MRI	MRI-CGCM3
23	Non-hydrostatic Icosahedral Atmospheric Model Group	NICAM	NICAM.09
24	Norwegian Climate Centre	NCC	NorESM1-M

[a]Two models capable of simulating the regional features over South Asia were used for the present study

(Sperber et al. 2013). Of them, the latest versions of the UK Meteorological Office model HadGEM2_CC and GFDL_CM3 developed at Geophysical Fluid Dynamics Laboratory, NOAA, Princeton, USA, can be used to get the future projections for the major rice growing states in India. The simulations from these CMIP5 global models are performed at a higher horizontal and vertical resolution compared to the CMIP3 groups, an added value for understanding and assessing regional changes in a warmer climate.

Monthly data for GFDL_CM3 and HadGEM2_CC for each of these GCMs from CMIP5 website was downloaded. The major climate variables required for the study included rainfall, maximum and minimum temperatures. We have selected the two GCMs (GFDL_CM3 and HadGEM2_CC) which have good skill in simulating realistic characteristics of Asian monsoons (Annamalai et al. 2013; Cherchi et al. 2014), and those models have better skill for simulating the future climate change scenarios.

The present approach considered the two selected global climate models (GCMs), and rainfall and temperature climatologies constructed from the past 30 years of the historical model simulations have been compared with observations. The robustness has been assessed based on the global climate models ability in representing the current climate regional features of our study regions. Hence, these models are considered for their capability in capturing the future projections in the study regions.

3.2 Approach and Methodology for Extracting Climate Data

The current and future climate change scenarios of rainfall, minimum and maximum temperature changes were generated for the 13 major rice growing states of India from the two selected GCMs. Figure 3.1 represents the schematic illustration for the downscaling approach adopted to prepare current and climate change scenarios for major rice growing states in India.

3.3 Spatial Domains of Major Rice Growing States in India

The demarcated spatial domain for the present study for the 13 states is given in Table 3.2. The larger domain will enable to ensure climatological consistency for the particular state which is used for impact studies. The monthly climate data, viz. rainfall, maximum and minimum temperatures for the respective states, was extracted for the current and future periods from the two different global climate models using the domain given below.

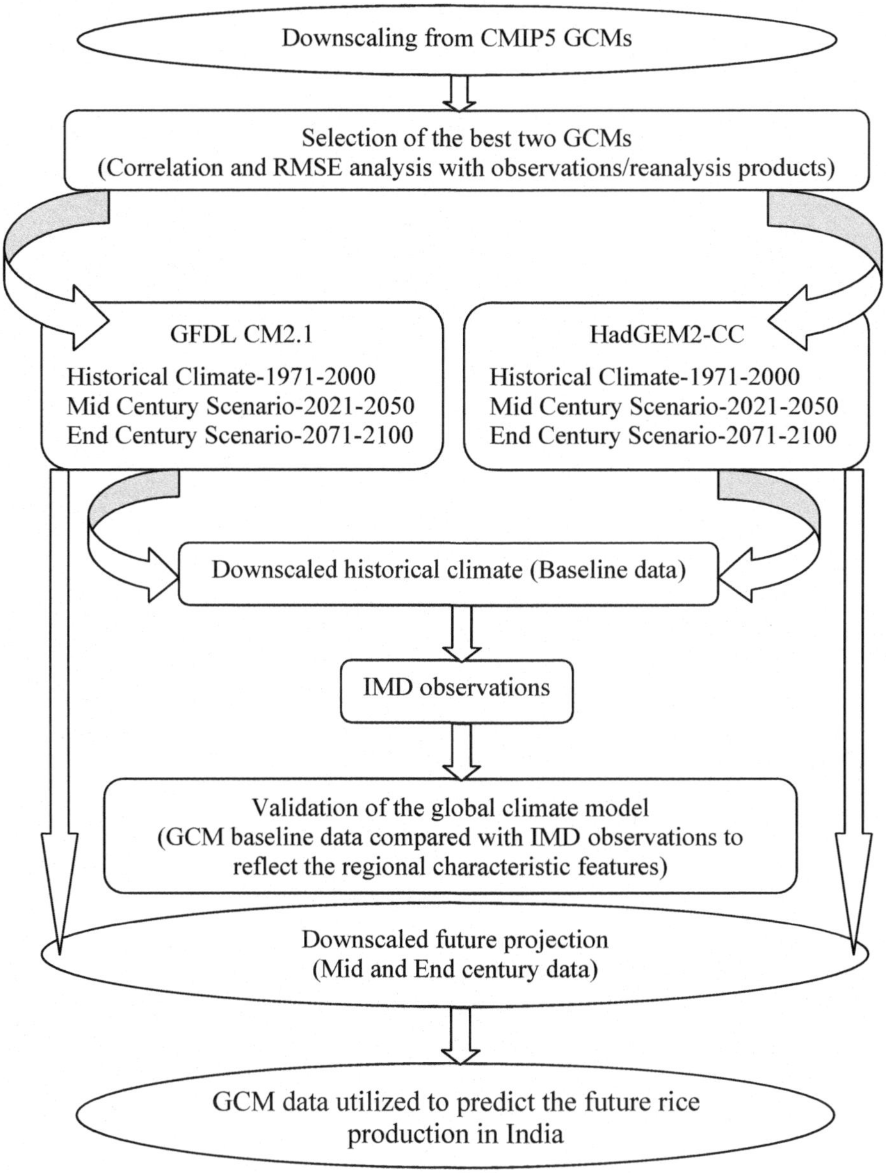

Fig. 3.1 Schematic diagram for extracting current and future climate change scenarios' data generation

Table 3.2 Spatial domain used for the selected states to extract climate data

S. No.	Name of the states	Longitude (°)	Latitude (°)
1	Andhra Pradesh	77–84.4	13–20
2	Bihar	83.3–88.2	24.5–27.5
3	Chhattisgarh	80.2–84.2	17.8–24
4	Gujarat	68.3–74.6	20–24.7
5	Haryana	74.5–78	27.7–31
6	Karnataka	74.2–78.6	11.5–18.5
7	Madhya Pradesh	74.1–82.7	21.1–26.6
8	Maharashtra	72.7–80.7	15.5–22
9	Odisha	81.8–87.5	18.0–22.5
10	Punjab	73.9–76.8	29.7–32.4
11	Tamil Nadu	76.5–80.3	8–13.6
12	Uttar Pradesh	77.2–84.6	24–30.5
13	West Bengal	85.8–89.1	21.5–26.5

3.4 Historical Observed Climate Data

Climate data relevant to the study area was collected from meteorological stations located in the respective states. These observational data was collected from the network stations maintained by India Meteorological Department (IMD), agricultural universities and State Department of Agriculture. We have taken observational data having long time series of nearly 30 years that is suitable for climatological trend analysis.

3.4.1 Seasonality of Rainfall

Climate normal of a variable is represented as an average value, typically for a period of 30 years. The seasonality can be expressed as the temporal evolution of rainfall which is capable of illustrating the timing of the maximum and minimum during a year. This is an effective indicator of climate of a particular state. The seasonal cycle of rainfall from observations for major rice growing states in India is shown in Fig. 3.2. It could be observed from Fig. 3.2 that rainfall peak is clearly observed during the south-west monsoon season from June to September in all the states except Tamil Nadu in which north-east monsoon is observed from October to December as a major monsoon season. This indicates that the clear climatological feature has been expressed by all the states from the observations. However, Karnataka receives the maximum quantity of rainfall, and Uttar Pradesh receives the lowest rainfall among the selected states.

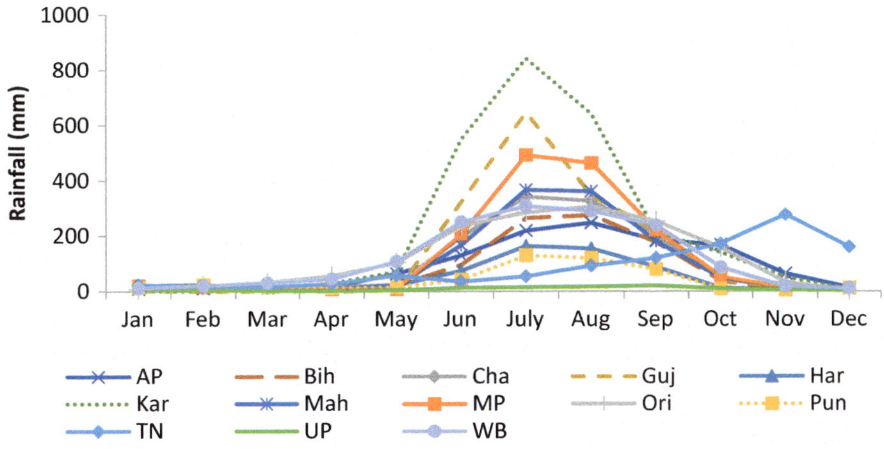

Fig. 3.2 Seasonal cycle of rainfall in major rice growing states of India

3.4.2 Seasonality of Temperature

The seasonal cycle of maximum and minimum temperatures from observations for major rice growing states in India is shown in Fig. 3.3. The highest values of maximum and minimum temperatures are encountered during the summer months of April–May, while the lowest values are observed during winter months December—January. This indicates that the maximum and minimum temperatures clearly brought out by the seasonal evolution in all the states.

3.5 Evaluation of Baseline Climate with IMD Observations

For all climate models, one of the most important challenging aspects is the simulation of rainfall climatology during monsoon seasons, viz. summer (June–September) and winter (October–December) monsoon periods. A realistic simulation of the basic state of monsoon rainfall climatology is a key feature, for assessing the future changes due to anthropogenic forcing over the Indian monsoon region. The rainfall extracted from the GCMs has been validated against observed rainfall climatology for all the major rice growing states of India. Compared to CMIP3, higher horizontal resolutions employed in CMIP5 models capture the topographically induced rainfall over the monsoon region. Therefore, it is useful to compare the seasonal cycles of rainfall and temperatures over the particular state. Comparing observations with model baseline data will provide a good indicator of the model performance to simulate climate change projections.

The selected CMIP5 models were evaluated for their ability to simulate precipitation and temperature patterns in the respective rice growing states in India. The

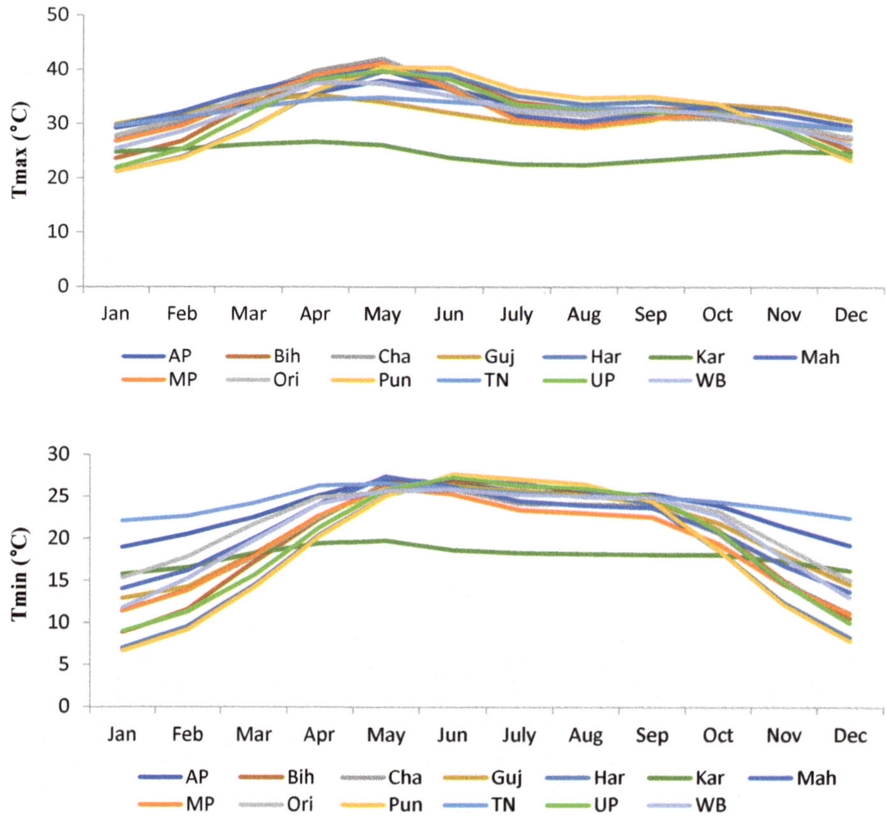

Fig. 3.3 Seasonal cycle of temperature in major rice growing states of India

baseline period selected for the study regions was from 1971 to 2000. The future time horizons up to 2100 have been analysed for this study for two future scenarios, viz. RCP4.5 and RCP8.5, representing medium- and high-emission futures. The future scenario selected for the period representing the mid- and end-century is, respectively, from 2021 to 2050 and 2071 to 2100 (Table 3.3).

3.5.1 Rainfall Seasonal Cycle

Seasonal cycle of rainfall for major rice growing states in India is given in Fig. 3.4. In all the states, the highest monthly normal rainfall was received during the south-west monsoon (June–September) season except Tamil Nadu. The peculiarity of Tamil Nadu rainfall seasonal cycle which is dominating influence of north-east monsoon is clearly seen during the months from October to December (Fig. 3.4k). The states

Table 3.3 Observational, baseline and future scenarios climate model data sets used in the study regions

S. No.	Product	Parameters	Scenarios	Time period	Frequency
1	Observed climate data from IMD stations	Rainfall (mm) Maximum temperature (°C) Minimum temperature (°C)	Station observatory located in different places	1971–2000	Monthly
2	GFDL-CM3	Rainfall (mm) Maximum temperature (°C) Minimum temperature (°C)	Baseline	1971–2000	Monthly
			Projection RCP4.5 and RCP8.5	2021–2050 and 2071–2100	
3	HadGEM2-CC	Rainfall (mm) Maximum temperature (°C) Minimum temperature (°C)	Baseline	1971–2000	Monthly
			Projection RCP4.5 and RCP8.5	2021–2050 and 2071–2100	

like Bihar, Chhattisgarh, Haryana, Odisha, Punjab, Tamil Nadu and West Bengal indicated that both the model rainfall has very closer agreement with the observations. The climate model underestimates the rainfall as compared to observations in the states like Gujarat, Karnataka, Madhya Pradesh and Maharashtra. This indicates that the degree of uncertainty prevails in rainfall observations. However, both the climate models predict the seasonal peak and magnitude of rainfall very well in all the states compared with observations.

3.5.2 Temperature Seasonal Cycle

Maximum temperature seasonal cycle for major rice growing states in India is shown in Fig. 3.5. It is observed that both the climate models capture the seasonal evolution of maximum temperature with slight cold bias in almost all the states except Karnataka which shows warm bias. It is observed that the monthly temperature evolution in all the states of baseline temperature has good similarity with the IMD observations. However, the magnitude of the maximum temperature is smaller than the rainfall. The states like Chhattisgarh, Madhya Pradesh, Odisha, Andhra Pradesh and Maharashtra have very close agreement between the model and observations compared to other states.

Seasonal cycle of minimum temperature for major rice growing states in India is given in Fig. 3.6. It is observed that both the climate models capture the timing of the highest and lowest temperatures in annual seasonal cycle thus confirming the good indicator of the model performance. In all the states, model minimum temperature agrees well with the observations.

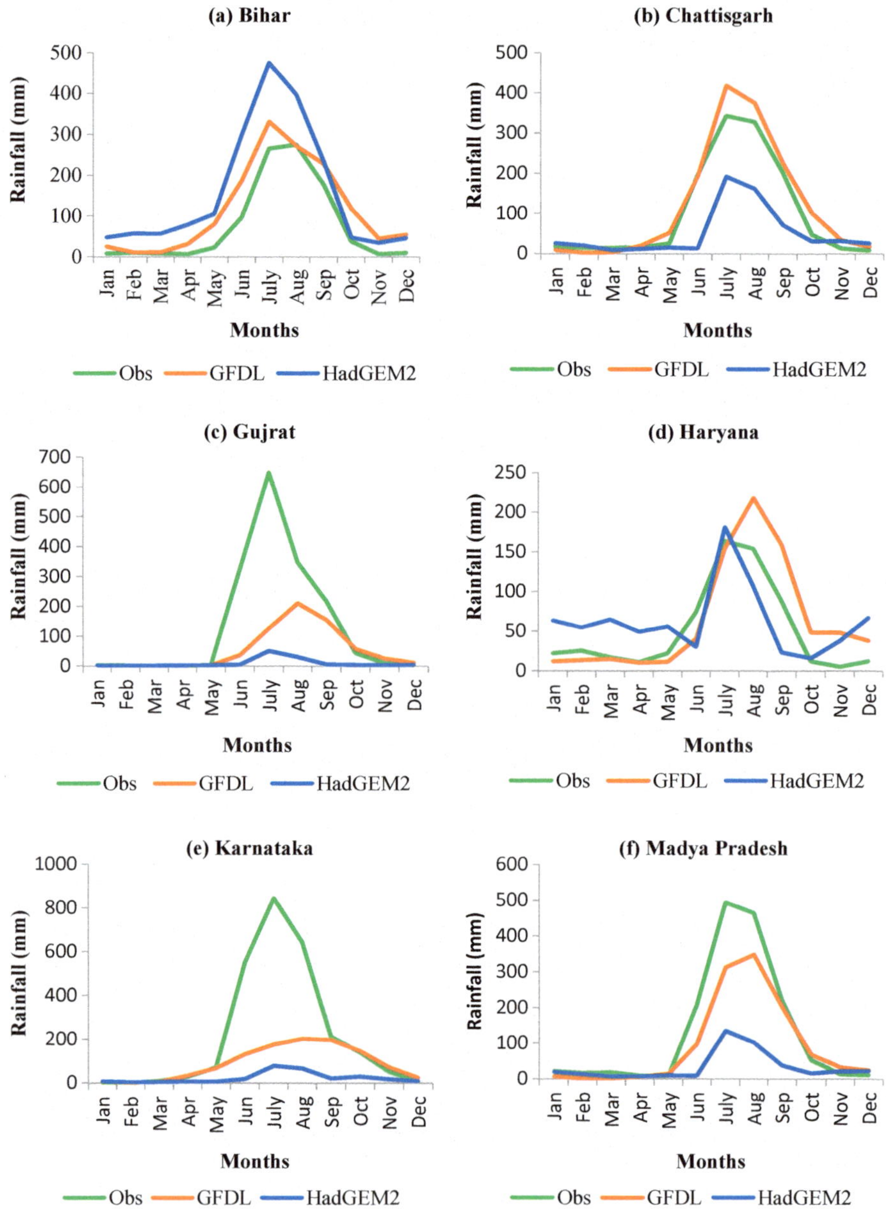

Fig. 3.4 Seasonal cycle of rainfall for major rice growing states in India: Evaluation of GCM (GFDL_CM3 and HadGEM2_CC) baseline rainfall with observations

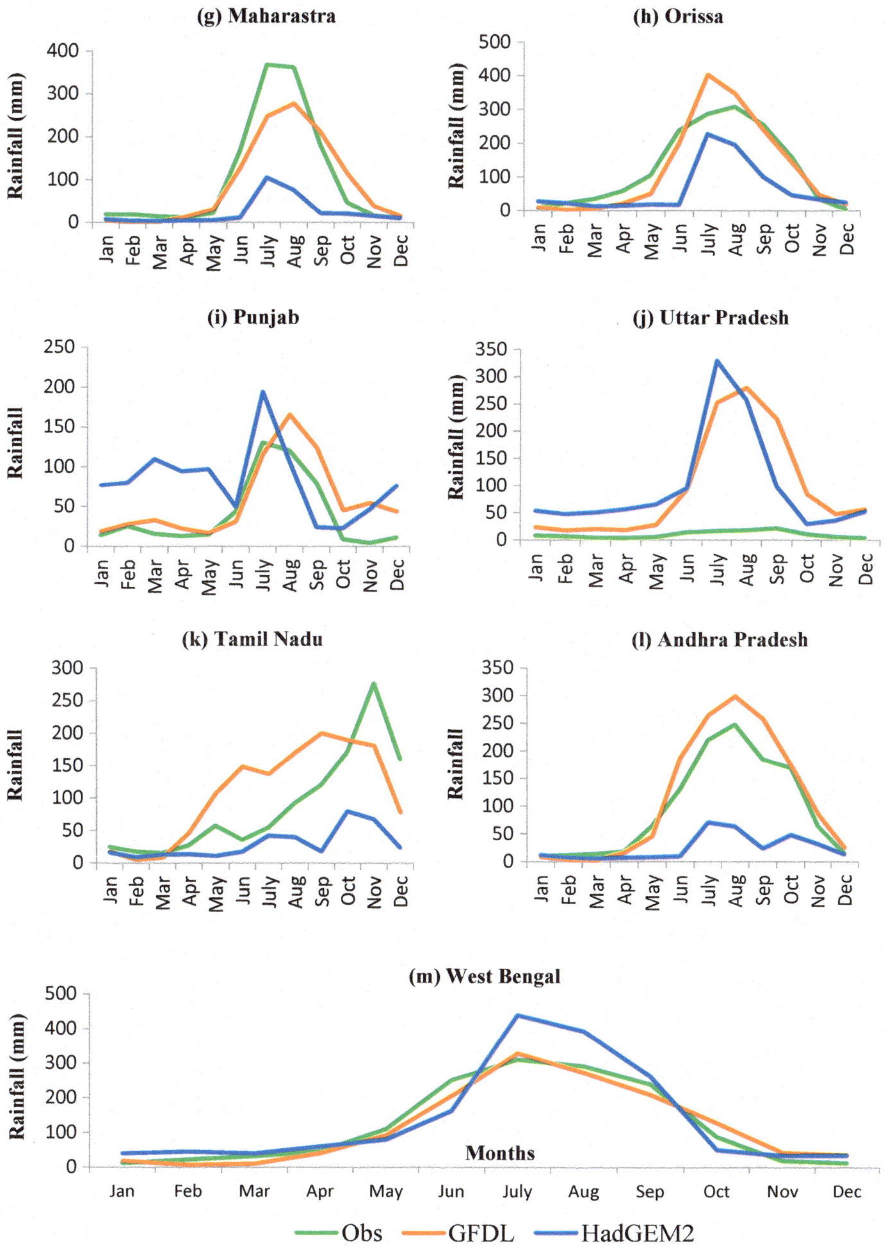

Fig. 3.4 (continued)

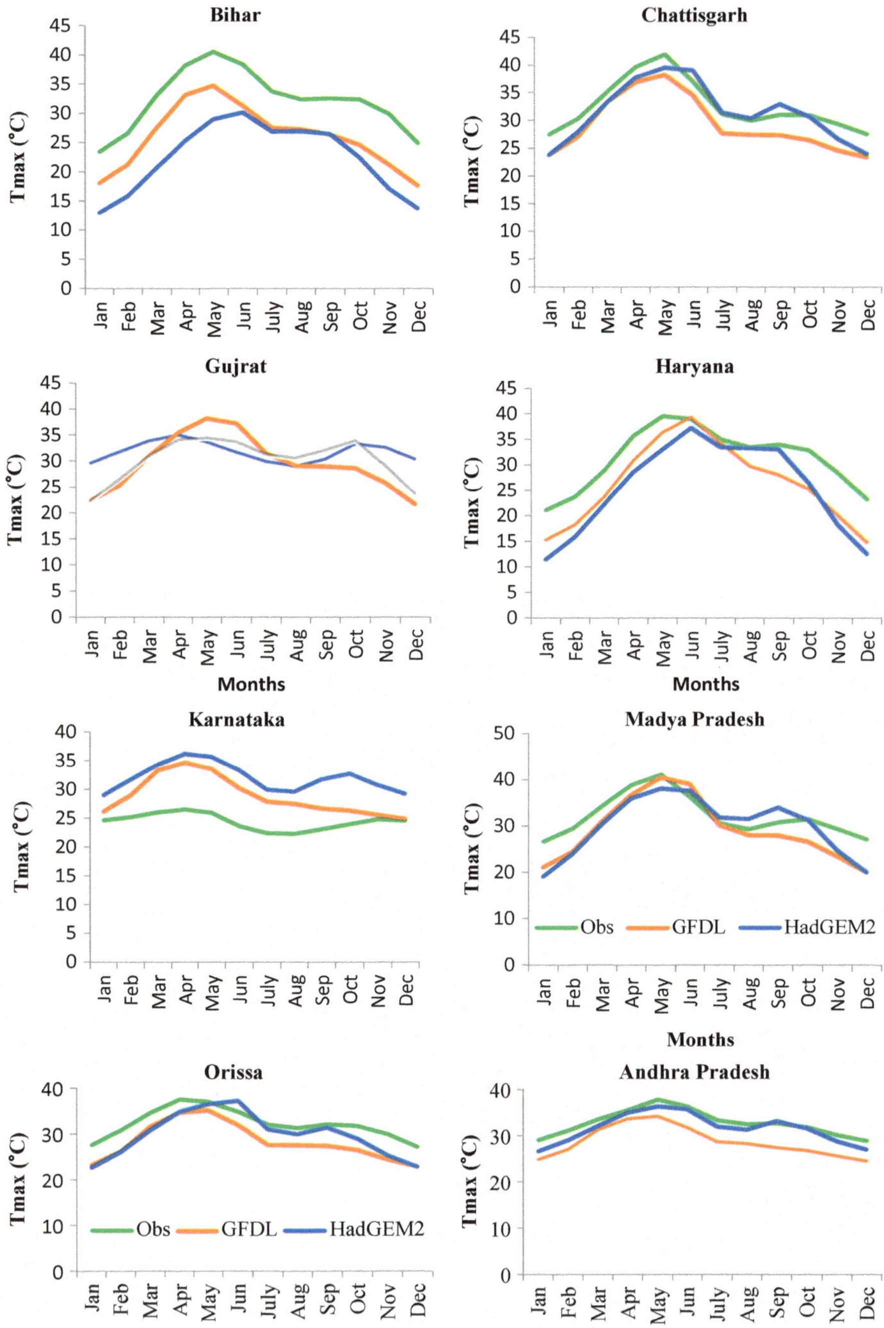

Fig. 3.5 Seasonal cycle of maximum temperature for major rice growing states in India: Evaluation of GCM (GFDL_CM3 and HadGEM2_CC) baseline with observations

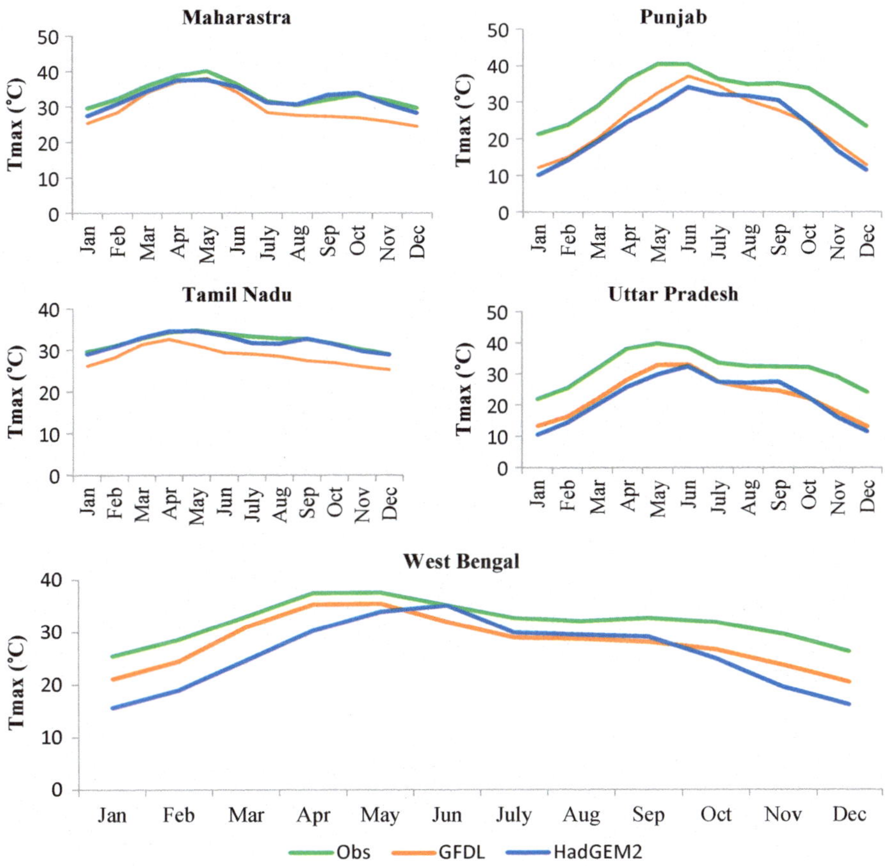

Fig. 3.5 (continued)

3.6 Future Climate Change Projections

The detailed future climate change projections were generated for the major 13 states from the two GCMs in the CMIP5. A new set of emission scenarios called the Representative Concentration Pathways (RCPs) was introduced in CMIP5. A particular RCP represents radiative forcing and CO_2 concentration reached by the year 2100, without being linked to any specific socio-economic development storylines as in the case of the earlier Special Report on Emission Scenarios (SRES). The CMIP5 in their Assessment Report 5 (AR5) models used the four RCP scenarios such as RCP2.6, RCP4.5, RCP6.0 and RCP8.5 in their future climate projection model runs.

For this study, the baseline and future time horizons up to 2100s for two future scenarios, viz. RCP4.5 and RCP8.5, representing medium- and high-emission scenarios were analysed. RCP4.5 represents the increases in radiative forcing (~4.5 W/m^2) and CO_2 equivalent (~650 p.p.m) up to 2100 and stabilizes after 2100 without overshoot

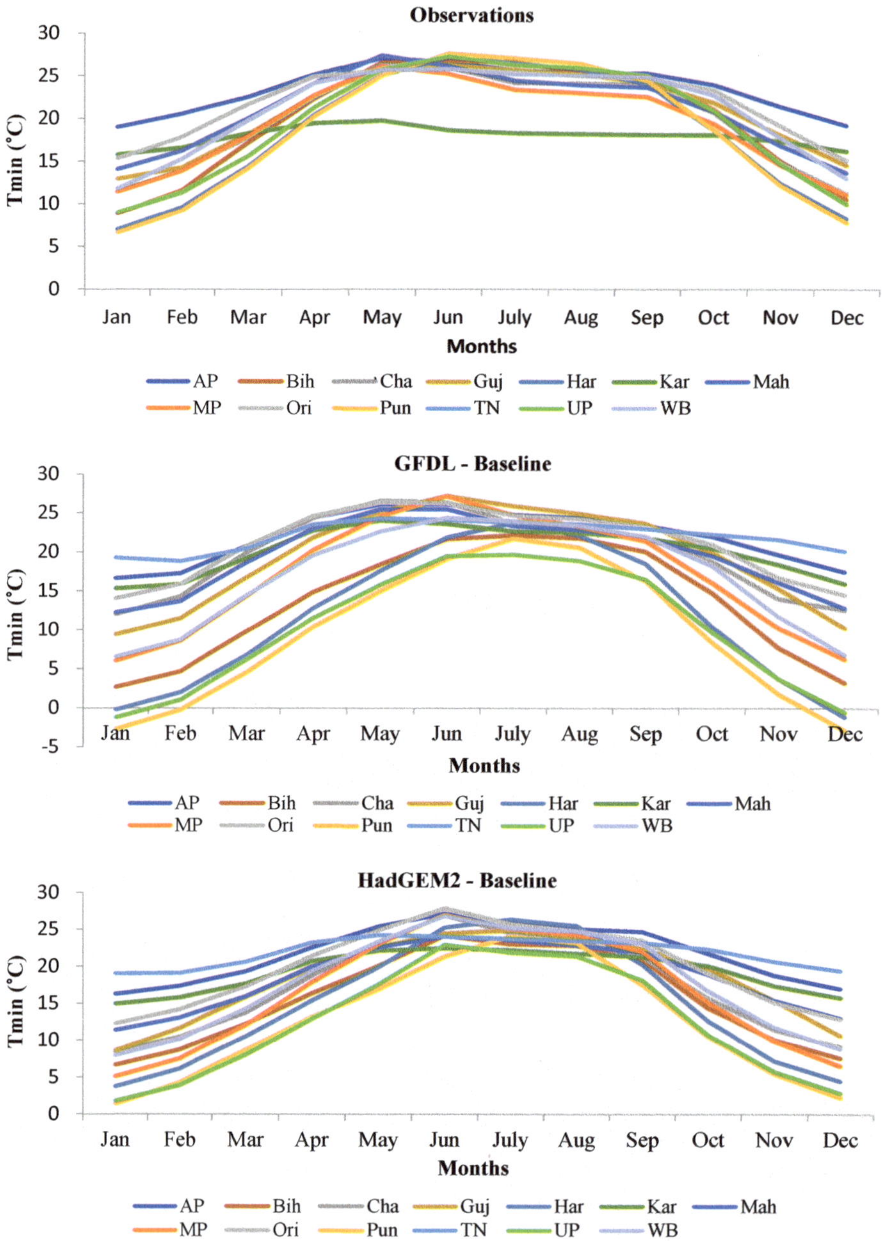

Fig. 3.6 Seasonal cycle of minimum temperature for major rice growing states in India: Evaluation of GCM (GFDL_CM3 and HadGEM2_CC) baseline with observations

relative to preindustrial conditions. Similarly, RCP8.5 represents the increases in radiative forcing (>8.5 W/m^2) and CO_2 equivalent (>1370 p.p.m) in 2100 and rising even after 2100 (Moss et al. 2010). The period 1971–2000 was defined as the baseline period. Then two time periods to represent the future, viz. mid-century (2021–2050) and end-century (2071–2100), were selected. Hence, the climate change projection was grouped under these two time periods and two emission scenarios for the present study.

3.6.1 Seasonal Cycle of Rainfall Projections of HadGEM2 Model

The rainfall climatology of baseline and future projections of HadGEM2 model were given in Fig. 3.7. The Hadley centre model predicts that the precipitation will increase throughout the year in all the states. Ratio of increase is prominent during the monsoon months compared to other months. Further, monthly rainfall climatology predicts the seasonal peak and trough shows a more confidence of the model projecting the future rainfall in all the states. Seasonal peak is noticed during the south-west monsoon season during June to September in almost all the states except Tamil Nadu with high north-east monsoon during October to December which is the major indicator of model projecting the future climate. For instance, Bihar rainfall increases about 7 and 22% under RCP4.5 and RCP8.5 scenarios, respectively, during the end of the century. In Tamil Nadu, both the scenarios during the mid-century show decrease in rainfall due to low radiative forcing. However, high anthropogenic force increases the rainfall about 3 and 27% under RCP4.5 and RCP8.5 scenarios, respectively, during end-century.

3.6.2 Seasonal Cycle of Maximum Temperature Projections of HadGEM2 Model

The maximum temperature climatology of baseline and future projections of HadGEM2 model were given in Fig. 3.8. The Hadley centre model projections indicate that the maximum temperature climatology shows an increasing tendency in all the states. It varies from 0.5 °C (Punjab) under RCP4.5 scenario during the mid-century and reaches the maximum increase of 6.1 °C (Gujarat) during the end-century under RCP8.5 scenario. For instance, Odisha maximum temperature increases about 1.07 and 1.17 °C under RCP4.5 scenario during the mid- and end-century, respectively. Further, it increases about 3.05 and 4.09 °C under RCP8.5 scenario during the mid- and end-century, respectively.

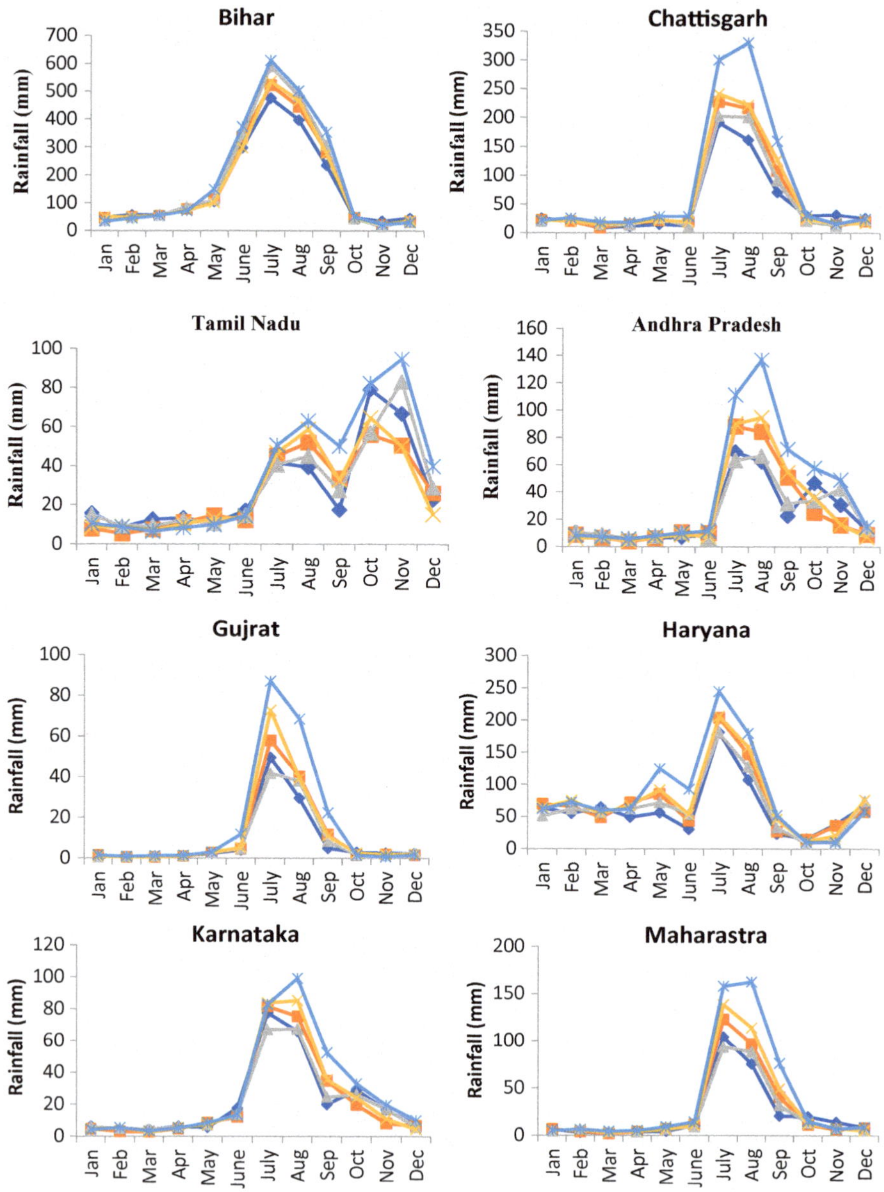

Fig. 3.7 HadGEM2 baseline and climate change projections of rainfall during 2050s and 2100s for two RCP scenarios

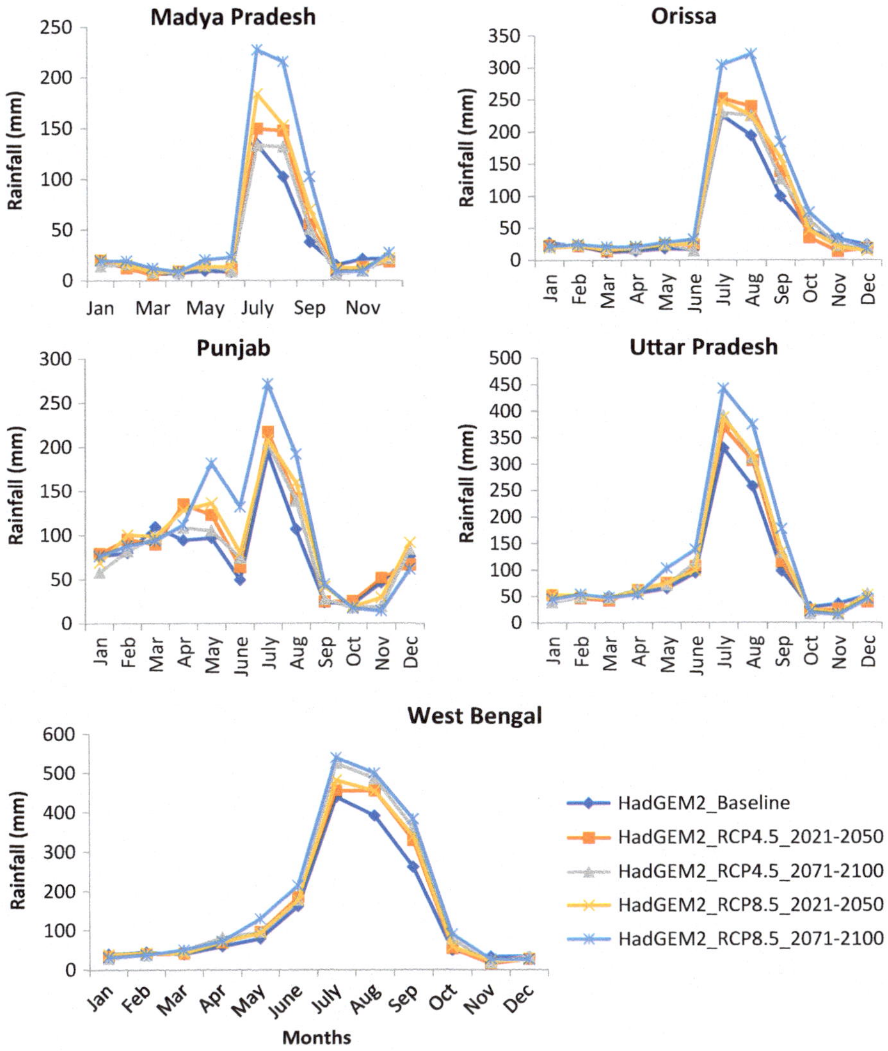

Fig. 3.7 (continued)

3.6.3 *Seasonal Cycle of Minimum Temperature Projections of HadGEM2 Model*

Baseline and climate change projections for minimum temperature of HadGEM2 model during 2021–2050 and 2071–2100s for two RCP scenarios are given in Fig. 3.9. The minimum temperature projections show increasing trend over the different states of India in almost all the months. The minimum temperature increase

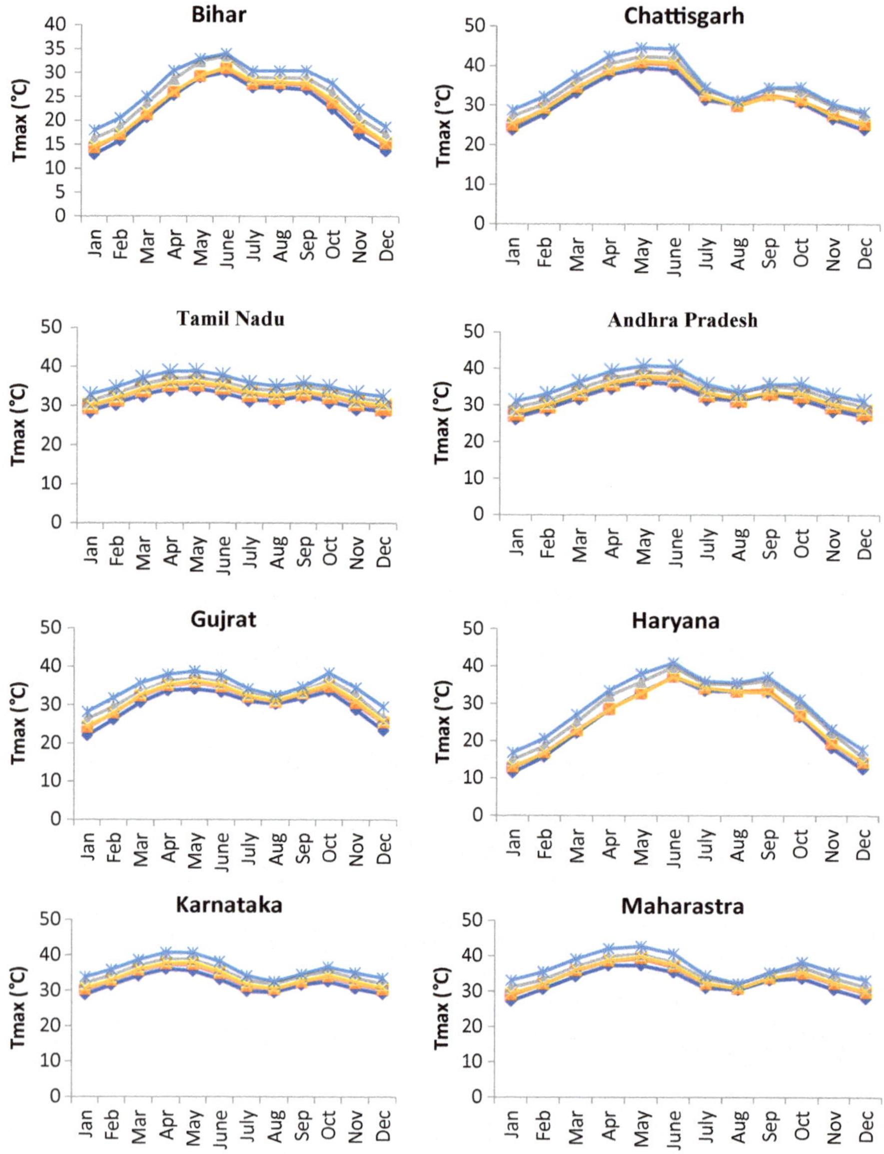

Fig. 3.8 HadGEM2 baseline and climate change projections for maximum temperature during 2050s and 2100s for two RCP scenarios

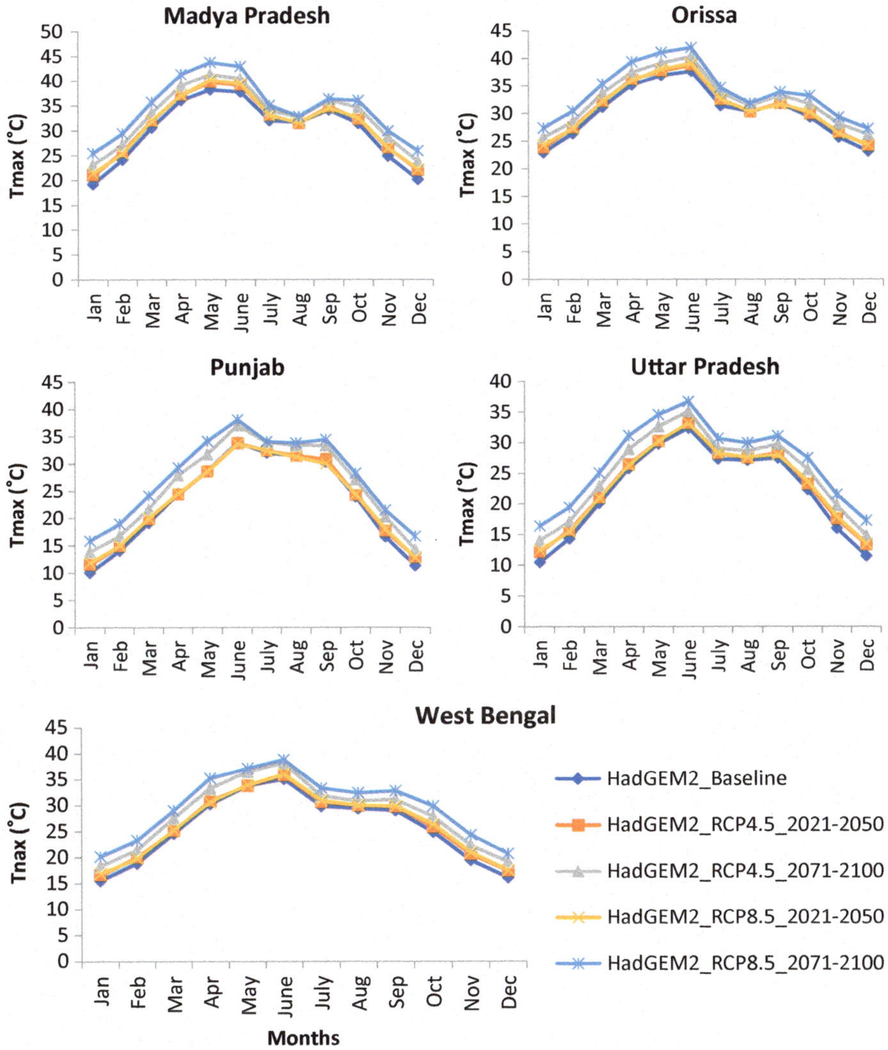

Fig. 3.8 (continued)

varies between 1.07 °C (Odisha) and 1.63 °C (Gujarat) under RCP4.5 scenarios during the mid-century. Further, it also varies between 4.55 °C (Andhra Pradesh) and 5.27 °C (Chhattisgarh) under RCP4.5 scenarios during the mid-century.

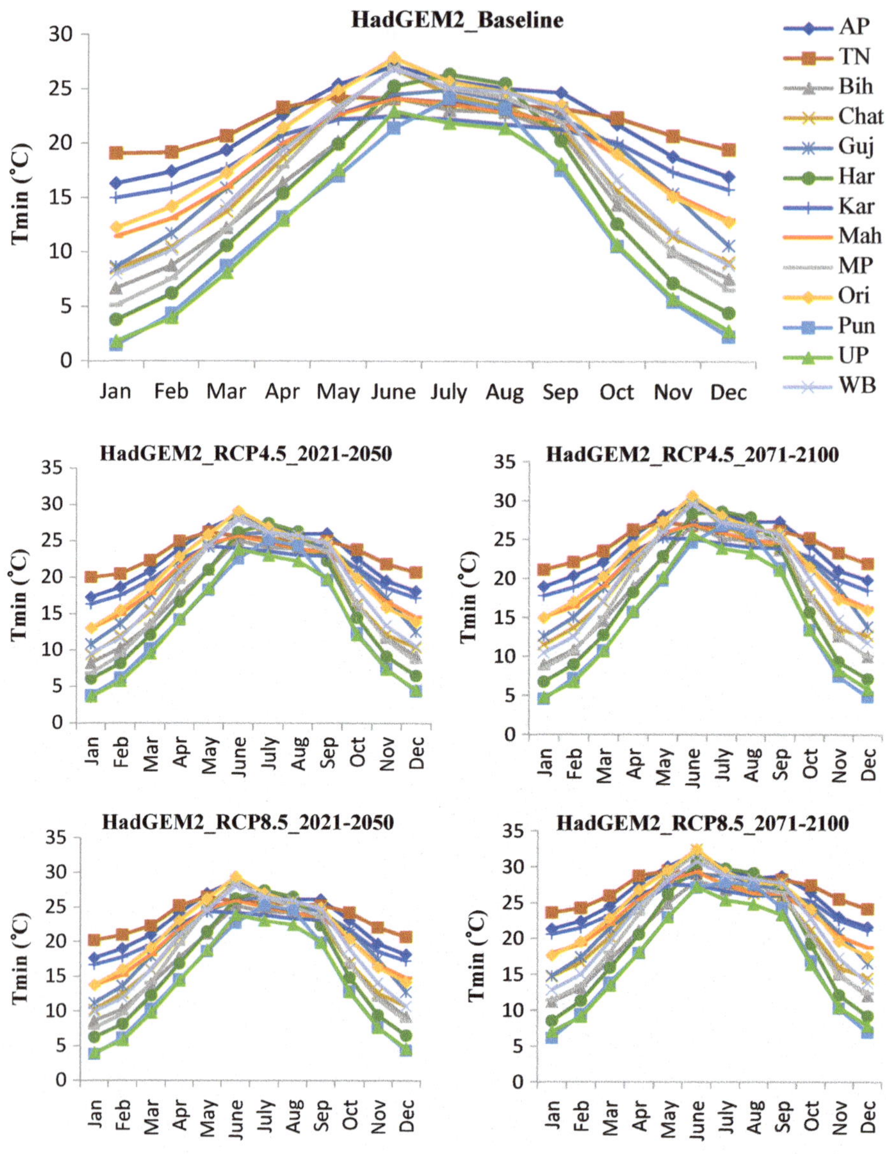

Fig. 3.9 HadGEM2 baseline and climate change projections for minimum temperature during 2050s and 2100s for two RCP scenarios

3.6.4 *Rainfall, Maximum and Minimum Temperature Projections of GFDL Model*

Baseline and climate change projections for annual rainfall, maximum and minimum temperatures of GFDL model during 2021–2050 and 2071–2100s for two RCP scenarios are given in Fig. 3.10. The GFDL model predicts that the annual precipitation will increase in all the states. The maximum and minimum temperature projections show increasing trend over all the states of India. The minimum temperature increase will be up to 2.26 °C in Haryana and 4.26 °C in Uttar Pradesh during the mid- and end-century, respectively, under RCP4.5 scenarios. The GFDL model projections indicate that the maximum temperature will increase up to 1.93 and 6.23 °C in Uttar Pradesh during the mid- and end-century, respectively, under RCP8.5 scenarios.

3.7 Summary

Future climate change projections for the major rice growing states are presented which help to predict the future rice production in India. Major focus is given for the selection of global climate modelling approach and methodology for extracting climate data, spatial domains of major rice growing states in India, historical observed climate data, evaluation of baseline climate with IMD observations and future climate change projections.

The two GCMs have good skill in simulating realistic characteristics of Asian monsoons, and those models have better skill for simulating the future climate change scenarios. The monthly climate data, viz. rainfall, maximum and minimum temperatures for the respective states, was extracted for the current and future periods from the two different global climate models using the demarcated spatial domain. Climate data relevant to the study area was collected from meteorological stations located in the respective states. The rainfall peak is clearly observed during the south-west monsoon season from June to September in all the states except Tamil Nadu in which north-east monsoon from October to December is observed as a major monsoon season. However, Karnataka receives the maximum quantity of rainfall, and Uttar Pradesh receives the lowest rainfall among the selected states. The highest values of maximum and minimum temperatures are encountered during the summer months of April–May, while the lowest values are observed during winter months December—January. The rainfall extracted from the GCMs has been validated against observed rainfall climatology for all the major rice growing states of India.

The selected CMIP5 models were evaluated for their ability to simulate precipitation and temperature patterns in the respective rice growing states in India. It was observed that the monthly temperature evolution in all the states of baseline temperature has good similarity with the IMD observations. It was also observed that both the climate models capture the timing of the highest and lowest temperatures in annual seasonal cycle showing the good indicator of the model performance. The

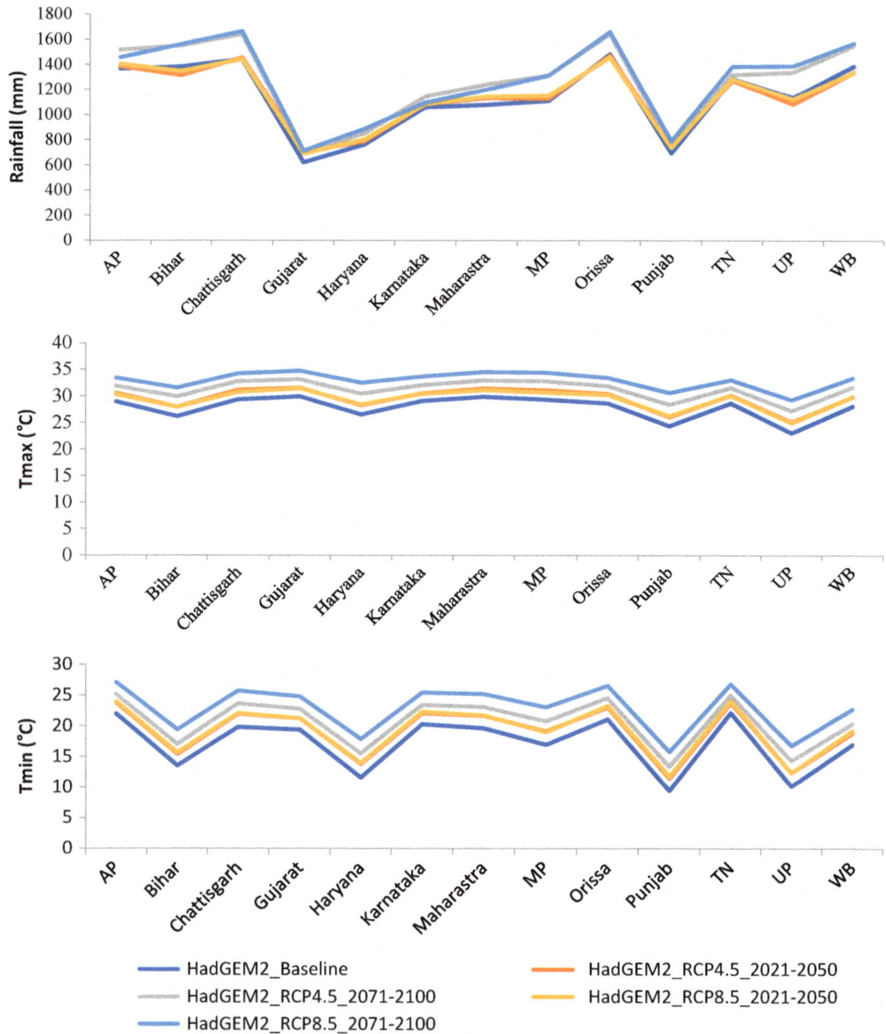

Fig. 3.10 GFDL baseline and climate change projections for annual rainfall, minimum and maximum temperatures during 2050s and 2100s for two RCP scenarios

detailed future climate change projections were generated for the major 13 states from the two GCMs in the CMIP5.

The baseline and future time horizons were analysed up to 2100s for two future scenarios, viz. RCP4.5 and RCP8.5, representing medium- and high-emission scenarios. Seasonal peak is noticed during the south-west monsoon season during June to September in almost all the states except Tamil Nadu which dominates north-east monsoon during October to December, thus showing the major indicator of model projecting the future climate.

Chapter 4
Methodologies for Quantifying Climate Change Impacts on Rice Production

K. Palanisami, Krishna Reddy Kakumanu, Udaya Sekhar Nagothu and C. R. Ranganathan

4.1 Introduction

In order to study how climate change affects agriculture, scientists normally use farm survey data collected by them, viz. primary data. They also use data from various other sources such as government publications, called secondary data. The data together is used in econometric models to get an understanding of the impact of climate change on the economy. Farm survey data usually results in cross-sectional data on farm characteristics. Farmers may suffer crop and income losses due to climate change impacts like droughts, floods and diseases to crops. Quantification of these losses is important but inadequately done due to lack of adequate data and relevant measurement methodologies. Hence, results of analysis of farm surveys data help us to quantify economic loss/gain, and this information can be used in short-term and long-term planning. Also using farm survey questionnaires, researchers can elicit information from the farmers about their opinion on climate change, impact of climate change shocks on agriculture and also the adaptation and mitigation strategies followed by them to control the negative effect of climate change on agriculture.

Long-term effects of climate change, from a macro-perspective, can be well studied from time series or panel data on yield of crops and climate-related variables. Econometric modelling of these data sets and using climate change projections will help to predict the possible increase/decrease in crop yield in future years. The information so derived can be used by policymakers for optimal cropping pattern.

Various statistical and econometric tools starting from simple averages to advanced econometric models are available at the disposal of the climate change modeller to understand how yield of crop is affected. These procedures help to formulate appropriate adaptation strategies which can be followed by government. In this chapter, various analytical tools used in assessing the influence of climate change on rice yield are presented.

K. Palanisami et al., *Climate Change and Future Rice Production in India*, India Studies in Business and Economics,
https://doi.org/10.1007/978-981-13-8363-2_4

4.2 Study Area and Data

For the present study, 13 Indian states where rice is predominantly grown were selected and grouped into five regions: southern, central, eastern, northern and western. From each state, one district which performed best in terms of productivity of rice for the past 30 or more years was identified. The districts selected from each state and the sample sizes from each district are given in Table 4.1 and Fig. 4.1. The study used both cross-sectional and panel data collected from the chosen districts. Using the cross-sectional data collected from farmers, the awareness of farmers on climate change and its effects were examined, while the panel data was used to predict long-term effects of climate variables on rice yield.

Cross-sectional data is collected from farmers through questionnaire. The survey was undertaken in 2014, and following information were collected from the famers.

(i) General information on the study area which includes the names of district/block/village, date of interview and names of the enumerators.

Table 4.1 Study area and the sample size

Sl. No.	State	District selected	Sample size (no. of farmers)
Region: Southern			
1	Tamil Nadu	Tanjore	112
2	Andhra Pradesh	West Godavari	101
3	Karnataka	Shimoga	100
Region: Central			
4	Madhya Pradesh	Balaghat	70
5	Maharashtra	Bhandara	100
Region: Eastern			
6	Bihar	Jehanabad	100
7	Odisha	Bhadrak	500
8	Chhattisgarh	Rajpur	70
9	West Bengal	Bankura	151
Region: Western			
10	Gujarat	Navsari	103
Region: Northern			
11	Punjab	Sangrur	100
12	Haryana	Karnal	100
13	Uttar Pradesh	Chandauli	100
		Total	1707

Note The sample sizes vary from 70 to 151 in all the states except for Odisha. This variation is due to availability of complete farm-level information. For Odisha, the sample size was 500 and it was drawn from a major research study conducted in that state

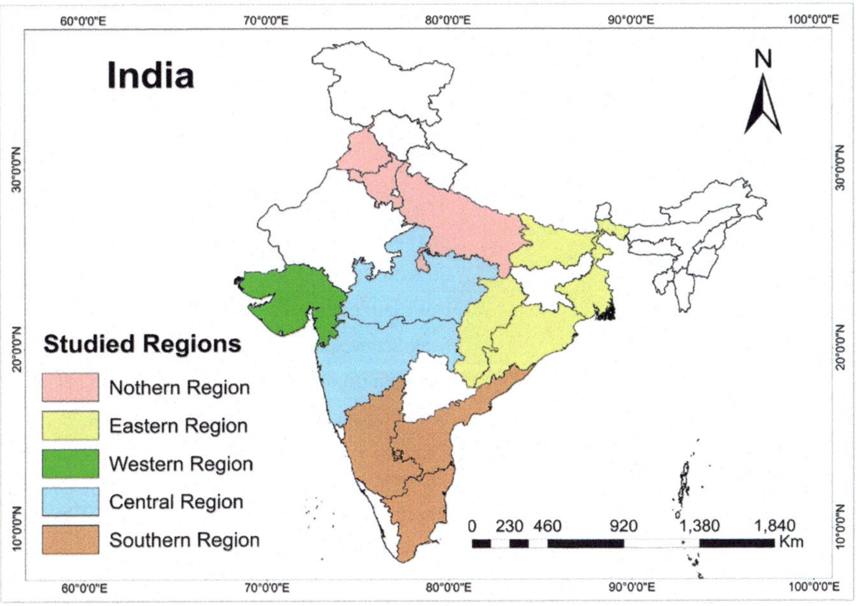

Fig. 4.1 Study regions

(ii) Socio-economic profile of the farm household. These include age, gender, educational level, occupation, main source income and membership any farmers associations.

(iii) Farm information such as land owned, type and source of irrigation, investments on irrigation, area grown under rice in each season, net income.

(iv) Farmers were questioned on their perception of climate change. They were asked to rank 16 climate change events (Table 4.2).

(v) Further, farmers were questioned on their opinion about the effects of climate change shocks (Table 4.3). They were asked to rank the following effects.

(vi) Farmers were asked to state the rice technologies or practices they adopt and the constraints if any in adopting. The crop management technologies or practices and non-crop-oriented strategies included were:

1. Crop management oriented:

    i. Direct sown rice (DSR)
    ii. System of rice intensification (SRI)
    iii. Modified SRI (MSRI)
    iv. Machine transplanting (MT)
    v. Improved management practices (change in variety, planting dates, efficient fertilizer and water use) (IMP)
    vi. Irrigation at minimum depth (IMD)
    vii. Alternate wetting and drying (AWD)

Table 4.2 List of events included for studying farmers' perception of climate change

Sl. No.	Variable	Notation used
1	Drought	Drought
2	Hailstorm	Hailstorm
3	Flood	Flood
4	Animal disease	AD
5	Increase in pest damage	IPD
6	Illness of family members	IFM
7	Untimely rains	UR
8	Irregular weather	IW
9	Temperature fluctuation high	TFH
10	Temperature fluctuation low	TFL
11	Change in soil salinity and decrease/increase in soil moisture	SSDI
12	Major changes in crop pattern	CCP
13	Major changes in livestock asset	CLA
14	Major changes in farm investment (capital formation)	CFI
15	Crop failure	CF
16	Depletion in groundwater	DGW

Table 4.3 List of effects of climate change

Sl. No.	Variable	Notation used
1	Decline in crop yield	DCY
2	Loss of assets	LA
3	Loss of income	LI
4	Food insecurity/shortage	FIS
5	Death of livestock	DL
6	Decline in consumption	DC
7	Decline in health	DH

viii. Supplemental irrigation through groundwater (SITG)
ix. Integrated pest management (IPM)
x. Machine harvesting (MH)
xi. Application of more organic manure (AMOM)
xii. Drip irrigation (DI)
xiii. Subsurface drainage (SSD)

2. Non-crop-oriented strategies

   i. Borrowed money from relatives/others
   ii. Received assistance from government/NGOs

iii. Less food consumption
iv. Shifted to non-farm employment
v. Outmigration to cities
vi. Government initiations to tackle climate change impacts.

At a macro-level, panel data was collected from historical records and government publications. The panel data consisted of

a. area, production and productivity of rice crop in the selected district over years
b. monthly data on temperature and rainfall in the selected district over years.

4.3 Statistical Tools for Analysis of Primary Data

a. **Garrett Ranking technique**

In the present study, farmers were asked to rank a number of alternatives (or factors) regarding climate change events (as listed in Table 4.2) and we are interested in knowing which factor is most preferred by the farmers. For this purpose, the Garrett Ranking technique (Garrett and Woodworth 1971) was applied. For example, in a farm survey in Tamil Nadu (Palanisami et al. 2014), farmers were asked about changes in climate over the last ten years and they were asked to rank the factors according to their perception:

1. Incidence of drought
2. Amount of rainfall/moisture
3. Rainfall distribution/delays
4. Extreme weather events
5. Temperature changes
6. Growing season.

In the Garrett Ranking method, the farmers' rankings are first converted as

$$\text{Percent Position} = 100\left(R_{ij} - 0.5\right)/N_j$$

where

R_{ij} Rank given for the ith factor by the jth farmer, and
N_j Number of factors ranked by the jth respondent.

By referring to Garrett's table, the per cent positions are converted into scores. Then for each factor, the scores of each individual are added from which the average scores can be obtained.

The frequency of the factor rankings can be listed in a table. An example of such a table is given below (Table 4.4).

The average scores were obtained as shown in Table 4.5.

Table 4.4 Factors influencing farmers towards climate change

Factor	Number of farmers who gave rank						Total
	1	2	3	4	5	6	
Incidence of drought	15	14	25	30	41	55	180
Amount of rainfall/moisture	23	52	53	29	19	4	180
Rainfall distribution/delays	45	62	41	15	14	3	180
Extreme weather events	83	31	28	18	12	8	180
Temperature changes	6	13	24	58	53	26	180
Growing season	8	8	9	30	41	84	180
Garrett score	77	63	54	46	36	23	

Table 4.5 Garrett ranking: selection of factors

Factor	Rank						Total garrett score	Average score	Rank
	1	2	3	4	5	6			
Incidence of drought	1155	882	1350	1380	1476	1265	7508	41.7	5
Amount of rainfall/moisture	1771	3276	2862	1334	684	92	10,019	55.7	3
Rainfall distribution/delays	3465	3906	2214	690	504	69	10,848	60.3	2
Extreme weather events	6391	1953	1512	828	432	184	11,300	62.8	1
Temperature changes	462	819	1296	2668	1908	598	7751	43.1	4
Growing season	616	504	486	1380	1476	1932	6394	35.5	6

Table 4.5 shows that farmers perceive extreme events as the most important climate change factor followed by rainfall distribution/delays.

The same procedure was employed in the present study with events as given in Table 4.2.

(a) Friedman Rank Test

Nonparametric tests can be applied to test the consistency of the rankings. For example, to test whether the farmers were consistent in ranking the six factors stated above, Friedman's test was employed. It is a nonparametric test, and it resembles two-way analysis of variance used in parametric tests. It is used to test for differences between groups when the dependent variable being measured is ordinal, for example, ranks. Friedman Rank Test is used to determine whether c groups have been selected from populations having equal medians.

$$H_0 : m_1 = m_2 = m_3 = \cdots = m_c$$

against the alternative, H_1 : Not all m_j are equal ($j = 1,2,\ldots c$).

The Friedman test statistic is given by

$$F = \frac{12}{rc(c+1)} \sum_{j=1}^{c} R_j^2 - 3r(c+1)$$

where R_j = Sum of the ranks for the jth group.

This statistic follows a χ^2 distribution with number of degrees of freedom equal to $c - 1$.

Thus, for any chosen level of significance, α, the null hypothesis is rejected if the calculated value of the F statistic is greater than the upper-tail critical value for the chi-square distribution with $c - 1$ degrees of freedom.

The results of the test, when applied to the above data set, are as given in Table 4.6.

In the above example, the test statistics has 5 degrees of freedom and the calculated value is 310.952. The table value of χ^2 at 5% significance level is 12.59, and so the null hypothesis is rejected which implies that the farmers did not consistently rank the factors; thus, the farmers' perceptions of climate change effects differ across the sample.

(b) **Kendall's coefficient of concordance**

Kendall's coefficient of concordance (Kendall and Smith 1939) is another statistical measure to test the consistency of rankings. This test statistic is defined by

$$W = \frac{12\,S}{m^2\left(M^3 - M\right)}$$

where m is the number of objects or factors (equal to 6 in the above example) and M is the number of judges (farmers = 180 in the above example) and S =

Table 4.6 Results of friedman rank test

Factor	R_j	R_j^2
Incidence of drought	773	597,529
Amount of rainfall/moisture	521	271,441
Rainfall distribution/delays	440	193,600
Extreme weather events	409	167,281
Temperature changes	757	573,049
Growing season	880	774,400
Total	3780	2,577,300
F statistic	310.95	
χ^2-table value (at 5 degrees of freedom)	12.59	

$\sum_{i=1}^{i=M}\left(R_i-\bar{R}\right)^2$; R_i = sum of the ranks of factor i and $\bar{R}=\frac{m(M+1)}{2}$. The quantity W is always positive but less than 1. If $W=1$, there is perfect agreement. But, if $W=0$ there is no overall agreement among the components in ranking the regions. To test the significance of W, use the statistic

$$\chi^2 = m(M-1)W$$

and it has a chi-square distribution with $M-1$ degrees of freedom.

For the data set discussed above, the results of Kendall's test are also in agreement with Friedman's test results. The Kendall's coefficient of concordance, W, was 0.346 which was found to be significant at 1% level implying no concordance or agreement in rankings.

4.4 A Review of Studies on Economic Impact of Climate Change on Crop Yields

There are many studies to estimate the impact of environmental changes on crop productivity. These studies used agro-economic models or multiple regression techniques. A brief review of studies done mostly after 2000 is presented below.

In agronomic–economic methods, crop models are used. They are developed from field experiments (FAO 2000; Kumar and Parikh 1998a). In these experiments, experiments are conducted in field or laboratory conditions with many valid future climates and carbon dioxide levels, but all farming practices in experimental conditions are held fixed. The aim is to elicit differences in outcomes which are due to temperature, precipitation or carbon dioxide. The cross-sectional approach is called as Ricardian method. In this method, the performances are studied across climate zones (Mendelsohn et al. 1994, 1996; Kumar and Parikh 1998b). This approach was used in studies across the USA (Mendelsohn et al. 1994, 1996)

Table 4.7 provides a selected list of papers on economic impact of CC on agriculture published during 2000–14. The major econometric model used in these studies is Just–Pope production function (Just–Pope 1978). The reason being it helps to estimate both mean yield and variability in yield simultaneously, and both these variables are affected by CC and so estimation of the effects of CC on these variables become easy.

Seo and Mendelsohn (2008) used structural Ricardian models and demonstrated that increases in temperature induce African farmers to adopt mixed farming and avoid specialized farms or livestock farms. Wang et al. (2009), in a study on China's agriculture, applied Ricardian modelling, and the crop net profit was the dependent variable and temperature and rainfall were the independent variables. Their study showed that temperature rise will affect rainfed farms, but it will be profitable to farmers in irrigated area.

Table 4.7 Selected list of papers on economic impacts of climate change published during 2000–2014

S. No.	Authors	Year	Country	Crop/variable	Technique used	CC variables
1	Mendelsohn et al.	2000	Africa	GDP	Simulation	IPCC forecast
2	Chen et al.	2004	US	Major US agricultural crops	JP	Average temperature and Annual Rainfall
3	Isik and Devadoss	2006	US	Wheat, Barley, Potato and Sugar beet	JP	Average temperature and Annual Rainfall
4	Schlenker and Roberts	2008	US	Corn, Soybeans and Cotton	Regression	Temperature data using Parameter -elevation Regressions on Independent Slopes Model (PRISM)
5	Seo and Mendelsohn	2008	Africa	Farming Income	Structural Ricardian Model	Summer, Winter temperature and Rainfall
6	McCarl et al.	2008	US	Corn, cotton, sorghum, soybean and winter wheat	JP	Average temperature, SD temperature and Annual Rainfall
7	Wang et al.	2009	China	Crop net revenue	Ricardian model	Temperature and Rainfall
8	Deressa and Hassen	2009	Ethiopia	Crop net revenue	Ricardian model	Seasonal Temperature and Rainfall
9	Kim and Pang	2009	Korea	Rice	JP	Average temperature and Annual Rainfall
10	Ranganathan	2009	India	Nine major crops grown in Tamil Nadu	JP	Average temperature and Annual Rainfall
11	Barnwal and Kotani	2010	India	Rice	JP and Qreg	Temperature, rainfall, SD Temperature and SD Rainfall
12	Cabas et al.	2010	Canada	Corn, soybean and winter wheat	JP	Grow days, temperature and rainfall

(continued)

Table 4.7 (continued)

S. No.	Authors	Year	Country	Crop/variable	Technique used	CC variables
13	Joshi et al.	2011	Nepal	Six major crops in Nepal	Time series analysis	
14	Rainer et al.	2011	China	Aggregate grain output from rice, wheat, corn, sorghum, millet, tubers and beans	JP	Annual average temperature, annual precipitation and average sunshine
15	Palanisami et al.	2011	India	Rice	JP	Average temperature and Annual Rainfall
16	Boubacar	2010	West Africa	Maize, Millet and Sorghum	JP	Standardized Precipitation Index (SPI), degree days, Average temperature and Annual Rainfall
17	Sarker et al. (a)	2012	Bangladesh	Rice	JP	Min and Max temperature, Rainfall
18	Sarker et al. (a)	2012	Bangladesh	Rice	JP	Min and Max temperature, Rainfall
19	Acquah et al.	2012	Ghana	Maize	JP	Average temperature and Annual Rainfall
20	Zainal et al.	2013	Malaysia	Crop net revenue	Ricardian model	Seasonal Temperature and Rainfall
21	Hasanthika et al.	2013	Sri Lanka	Rice	JP	Min and Max temperature, Rainfall
22	Pattanayak and Kavikumar	2013	India	Rice	Regression, simulation	Min and Max temperature, Solar Radiation, Rainfall
22	Mahadeb et al.	2014	Nepal	Rice, Maize and Wheat	JP	Temperature and annual Rainfall

In a study to understand the effect of CC on crop production in Ethiopia, Deressa and Hassen (2009) applied Ricardian modelling approach. Their study revealed that a small increase in temperature in summer and winter would significantly reduce crop net revenue per hectare. But increase in rainfall precipitation during spring would significantly increase net crop revenue per hectare. Similarly, Zainal et al. (2013) applied and analysed the CC impact on crop net revenue. The study indicated that temperature and rainfall had significant negative impact on production of paddy in the study area.

The agro-ecological zone-based approach was also used (AEZ) (FAO 1992, 1996). Variability in crop yields is important because agricultural production is very sensitive to changes in rainfall and temperature. Not much attention was paid to study this important aspect (Bindi et al. 1996; Mearns et al. 1997). Econometric models are usually applied to study the effect of climate change on impact of climate yield variability. Mendelsohn et al. (2000) used simulation of IPCC forecast to estimate the GDP loss due to CC in African agriculture. They found that African agriculture is very vulnerable to CC and that potential damage may be large both in absolute terms and as a fraction of agricultural GDP. Schlenker and Roberts (2008) studied the nonlinear temperature effects on crop yields. Chen et al. (2004) and Isik and Devadoss (2006) have applied regression technique by employing a Just–Pope (1978) production functional form. They estimated both the mean and variability in crop productivity. Chen et al. (2004) demonstrated that changes in climate variables modify crop yield values and variances.

Isik and Devadoss (2006) used Just–Pope production function to study the impacts of projected climate change on the yield of wheat, barley, potato and sugar beet in Idaho, USA. The independent variables employed were total precipitation, temperature and trend. They showed that climate change will have limited effects on the mean yields, but will significantly decrease the variance and covariance for most of the crops used in the study. Ranganathan (2009) used Just–Pope function methodology to estimate effects of climate change on nine major crops grown in Tamil Nadu. Further, by using multi-goal linear programming technique, he formulated optimal cropping plan for Tamil Nadu for 2021 and 2026. Palanisami et al. (2011) studied the CC impacts of agriculture in Godavari basin, Andhra Pradesh, India. The study revealed that rice production will decrease during mid- and end-century periods by 16 and 36%, respectively. But if water and labour-saving technologies are employed, the reduction in rice yield will be removed during mid-century and it will be only 19% during end-century period. Also if these technologies are followed, on the whole, water saving will be about 20%

Barnwal and Kotani's (2010) study revealed that increase in temperature as well as inter-annual variance of temperature and rainfall adversely affected the average yield of rice in Andhra Pradesh, India. Also, increase in average temperature, rainfall and their respective variances are likely to increase inter-annual variability in rice yield. Using quantile regression equations, they showed that changes of rice yield to climate differ significantly across the quantiles of the yield distribution. Sarker et al. (2012) studied the climate change effects on three major rice varieties: Aus, Aman and Boroin Bangladesh. They used Just–Pope function to estimate the mean and variability in yield. Their study revealed that the effects of climate change variables vary among the varieties. Kim and Pang (2009) analysed the climate change effects on rice yield in Korea by employing Just–Pope production function. They found that mean yield is positively related to temperature and negatively related to rainfall while both climate variables are risk increasing for variability in yield. Hasanthika et al. (2013), using Just–Pope production function, studied the climate variability, risk and paddy production. The study showed that climate factors, rainfall and maximum temperature, and production factors such as labour, machinery cost and cultivation extents are positively related to the probability distribution of yield. Therefore, their variability induces increase in risk.

In a study on crop yields and yield variability on maize, millet and sorghum in Sahel, Boubacar (2010) found that increase in degree days will be harmful to crop yields, but soil property index and precipitation intensity index statistically significant positive impacts in mean yields. Cabas et al. (2010) examined the effect of climatic and non-climatic factors on the mean and variance of corn, soybeans and winter wheat in southern Ontario, Canada, over a period of 26 years, 1981–2006, using Just–Pope modelling framework. They included input change, area change, monthly temperature and monthly rainfall from April to October as explanatory variables.

In Just–Pope modelling, an important assumption is stationarity of yield distribution. This hypothesis was examined by McCarl et al. (2008). They used temperature, precipitation, variance of intra-annual temperature, index of rainfall intensity and Palmer Drought Severity Index (PDSI) as the main variables in the analysis. They found that stationarity of yield distribution is not tenable when the mean and variance of key climate variables change over time. Also they showed that the mean of the crop yields is affected by the average temperature and precipitation. They concluded that stationarity is a doubtful assumption and stressed the importance of testing this assumption before going for probabilistic models.

One study on positive impact of climate change has been made by Holst et al. (2010). They used JP methodology and found that grain production in South China might be a beneficiary of global warming in the short run and estimated that 1 °C increase in annual average temperature in South China would increase the grain output by 3.79 million tons. In contrast to this, the study revealed that the impact of global warming in North China is negative, small and insignificant.

De-Graft Acquah, Henry and Kyei, Clement (2012) using JP framework studied the effect of CC on maize yield. The study showed that mean maize yield has positive relation related to crop area but has negative association with rainfall and temperature. The study also indicated that when crop area and temperature increase, maize yield variability also increases. But when precipitation decreases, variability in maize yield decreases.

4.5 Econometric Tools for Studying Climate Change Impacts

This section describes various statistical and econometric tools employed in the present study.

a. **Testing for normality of data**

Testing for normality of yield data is the first step in any econometric modelling as various tests of the parameters are based on this assumption especially if we want to apply Maximum Likelihood Estimate (MLE) methods. There are many tests available for this purpose. The important tests are (i) Shapiro–Wilk test and Lilliefors test. These tests have been applied for yield data prior to modelling. Both of these tests can be done using software packages such as Gretl.

b. **Just–Pope Production Function**

The centre of interest of the present study is on the impact of climate change on rice crop production in the five regions of India using an econometric approach. Specifically, we applied Just–Pope production function and quantile regression. The analysis is done separately for each region. The mathematical relation between productivity (y) of rice and a set of explanatory variables (X) denoted by the vector is given by the Just and Pope stochastic production function (Just and Pope 1978):

$$y = f(X; \beta) + \omega h(X; \delta)^{0.5} \tag{4.1}$$

where ω is the random term with mean zero and variance 1,β and δ are the constants to be estimated from past data. The error term,$\omega h(X; \delta)^{0.5}$ in Eq. (4.1), shows that Just–Pope model has heteroscedasticity error terms. The variable y is a panel data consisting of the yield per hectare observed over a period of time across the states belonging to the region. The set of explanatory variables include trend, average temperature, annual precipitation and dummy variables representing the states belonging to the region. The expected rice productivity is $E(y) = f(X; \beta)$, and crop variability is $V(y) = h(X; \delta)$. Therefore, $f(X; \beta)$ and $h(X; \delta)$ are called **mean** and **variance functions**, respectively. The derivatives of $h(X; \delta)$ w.r.t. precipitation and temperature can be used to check if a climate variable increases or decreases crop variability. Therefore, if, $h_x = \frac{\partial h}{\partial x} > 0$, it shows that the corresponding independent variable x will increase the risk. On the other hand, if $h_x < 0$, it means risk decreasing. This is the main advantage of using the Just–Pope production function. We can estimate both the mean yield and also variability in yield simultaneously.

The procedure to estimate the production function can be considered as estimation with heteroscedastic errors as in the following equation (Saha et al. 1997; Kumbhakar 1997)

$$y = f(X; \beta) + u \tag{4.2}$$

where $u = \omega h(X;\beta)^{0.5}$ with $E(u) = 0$ and $\mathrm{Var}(u) = h(X;\delta)$. There are two approaches suggested in many studies to estimate the mean and variance functions of the Just–Pope production function. They are Feasible Generalized Least Squares (FGLS) and Maximum Likelihood Estimation (MLE).

The two methods have their own advantages and disadvantages. The FGLS approach is employed in many studies. For large samples, FGLS approach is used to estimate fixed effects models (Judge et al. 1988). Further, our panel data consists of yield of rice across different districts observed over a period of about 30 years. So the estimation may lead to the problems of heteroscedasticity and autocorrelation (Gujarati 2004). As stated by Sarker et al. (2012), these two problems are better addressed in FGLS methodology. On the other, the MLE approach produces consistent and more efficient estimates for the parameters than FGLS approach and more suitable for small samples (Saha et al. 1997). The two methods are described below.

The essential steps for estimating the Just–Pope production function using FGLS are as follows:

1. Estimate the mean function by running OLS regression of y (yield of rice per ha) on $f(X;\beta)$.
2. Compute the residuals u_{it}.
3. Regress natural logarithm of u_{it}^2, i.e. $\log\left(u_{it}^2\right)$ on $\log(h(X;\delta))$, and get the residuals in the second regression.
4. Obtain the antilogarithms of the residuals obtained in Step 3. These are consistent estimates of the variances.
5. Again estimate the original mean function $f(X;\beta)$ by weighted least squares with inverse of variances (obtained in step 4) as weights. That is, we apply WLS to estimate the equation

$$yh^{-1/2}(X;\delta) = f(X;\beta)h^{-1/2}(X;\delta) + uh^{-1/2}(X;\delta). \tag{4.3}$$

In the MLE approach, the log-likelihood function is maximized. If the error term ω is assumed to be $N(0, 1)$, Eq. (4.1) shows that the variable y has a *normal distribution* with mean $f(X;\beta)$ and variance $h(X;\delta)$ and so the log-likelihood function can be written as

$$yh^{-1/2}(X;\delta) = f(X;\beta)h^{-1/2}(X;\delta) + uh^{-1/2}(X;\delta) \tag{4.4}$$

The parameters β and δ are estimated by maximizing the above function.

Our experience with data from farm survey shows that, in general, the optimum value of log-likelihood function under MLE is greater than that estimated by FGLS implying that MLE estimates are superior. But for some data, converge problems occur when estimation is done using MLE while no such difficulty arises under FGLS. Hence in the present study, both methods are employed. We first apply MLE and estimate the parameters and if iterations do not converge we use FGLS approach.

The two methods of estimation can be implemented using Gretl Econometrics package or STATA. The sample Gretl Script file and Do file of STATA are shown in Appendix.

The J-P production function was estimated by identifying the explanatory variables and then explicitly stating the forms of the mean and variance functions: $f(X;\beta)$ and $h(X;\delta)$. In the literature, for rice crop, different climate variables and different functional forms have been used (see Table 4.8 for details). Studies by Chen et al. (2004), Isik and Devadoss (2006), Ranganathan (2009), Palanisami et al. (2011), Kim and Pang (2009) have used a quadratic functional form for the mean function using average temperature during growing season and annual rainfall. For the variance function, either exponential or Cobb–Douglas form was used. Aye and Ater (2012) used linear, quadratic and square root forms for the mean function and linear form for the variance function. McCarl et al. (2008) introduced the standard deviation of temperature in the mean and variance functions to study the climate change effects on corn, cotton, sorghum soybeans and winter wheat. Similarly, Barnwal and Kotani (2010) included standard deviation of rainfall and temperature in the two functions because variation in temperature and rainfall adversely affects the mean and variance in rice yield. Recent studies show that rather than average temperature, maximum and minimum temperatures have stronger effects on rice yield. For example, Pattanayak

Table 4.8 Functional forms of Just–Pope function

Sl. No.	Variables	Functional forms[a] for	
		Mean	Variance
1	Trend, Avg*T* and Rain*T*	(i) Linear in trend, Avg*T* and Rain*T* (ii) Linear in trend and quadratic in Avg*T* and Rain*T* (iii) Linear in trend and Cobb–Douglas in AvgT and RainT (iv) Linear in trend and square root in AvgT and RainT	(i) Exponential w.r.t. trend, Avg*T* and Rain*T* (ii) Cobb–Douglas w.r.t. trend, Avg*T* and Rain*T*
2	Max*T*, Min*T*, Rain*T*, trend	(i) Linear in trend and quadratic in Max*T*, Min*T* and Rain*T* (ii) Linear in trend and Cobb–Douglas in Avg*T* and Rain*T*	Cobb–Douglas w.r.t. Max*T*, Min*T*, Rain*T* and trend
3	Avg*T*, Rain*T*, SD(AvgT), SD(RainT) and trend	Linear w.r.t. all variables	Linear w.r.t. all variables

Avg*T* = Average temperature; Rain*T* = Annual/seasonal rainfall; SD(T) = standard deviation of rainfall; SD(Rain*T*) = Standard deviation of rainfall; Max*T* = Maximum temperature; Min*T* = Minimum temperature and t = trend

[a]The functional forms are given omitting the dummy variable terms

and Kavi Kumar (2013) have shown that both minimum and maximum temperatures adversely affect rice yield during different growth phases. They have estimated an annual loss of 172 million tons. Similarly, Sarker et al. (2012) have studied the effect of changes in maximum and minimum temperatures on rice yield. They have both Cobb–Douglas and quadratic forms for the mean function and Cobb–Douglas form for the variance function. Similarly, Hasanthika et al. (2013) included these two variables to study the climate variability and risk in paddy production in Sri Lanka.

Thus, we find that while average temperature and rainfall are most commonly included in the mean and variance functions, use of maximum, minimum and standard deviation of temperature and rainfall are more plausible from the biological point of view in explaining the mean and variability in rice yield.

The above discussion shows that the two functions can be stated as:

Model 1

Mean Function:

Linear

$$f(X;\beta) = \beta_0 + \beta_1 t + \beta_2 T + \beta_3 P + \sum_{i=1}^{i=S-1} \beta_{6+i} D_i \tag{4.5a}$$

Quadratic

$$f(X;\beta) = \beta_0 + \beta_1 t + \beta_2 T + \beta_3 P + \beta_4 P^2 + \beta_5 T^2 + \beta_6 PT + \sum_{i=1}^{i=S-1} \beta_{6+i} D_i \tag{4.5b}$$

Square root

$$f(X;\beta) = \beta_0 + \beta_1 t + \beta_2 T + \beta_3 P + \beta_4 P^2 + \beta_5 T^2 + \beta_6 PT + \sum_{i=1}^{i=S-1} \beta_{6+i} D_i \tag{4.5c}$$

where t represents time trend, T is the average temperature during the rice growing months in a year, P is the annual rainfall, $D_i, i = 1, 2 \ldots S$ are the district dummy variables which take values 1 and 0 and S is the number of states included in the region. Annual rainfall not only reflects rainfall falling directly on the crop but also water accumulated between seasons within the year (Isik and Devadoss 2006). These forms of mean function are consistent with the postulates of Just and Pope (1978). Further, these forms are flexible enough to estimate using MLE or FGLS. The inclusion of time trend represents technological progress in agricultural front during the sample periods.

Variance Function

The variance function $h(X;\delta,\eta)$ was assumed to have exponential form

Exponential:

$$h(X;\delta) = \exp(\delta_0 + \delta_1 t + \delta_2 T + \delta_3 P) \tag{4.6a}$$

Cobb–Douglas:

$$h(X;\delta) = \beta_0 t^{\alpha} \prod_j x_j^{\beta_j} \tag{4.6b}$$

The exponential was employed first by Harvey (1976) and after that by many authors (Palanisami et al. 2011; Ranganathan 2009; Isik and Devadoss 2006; Isik and Khanna 2003; Asche and Tveteras 1999). The important benefit of this mathematical form is that it guarantees positive output variance. Also, the riskiness of an input variable can be quickly obtained by examining the sign of the coefficient of that variable in the function. For example, with the above functional form, it can be easily checked that $h(X;\delta) = \beta_0 t^{\alpha} \prod_j x_j^{\beta_j}$. As the mathematical function h is always positive, precipitation will be risk increasing if $\delta_1 > 0$ and it will be risk decreasing if $\delta_1 < 0$. In view of these advantages, in the present study, the exponential form of the variance function is employed.

Thus, Model 1 has three submodels:

(i) Linear mean function and exponential variance function
(ii) Quadratic mean function and exponential variance function and
(iii) Square-root mean function and exponential variance function.

Model 2
Mean Function
Linear:

$$f(X;\beta) = \beta_0 + \beta_1 t + \beta_2 \text{Max}T + \beta_3 \text{Min}T + \beta_4 \text{Rain}T + \sum_{i=1}^{S-1} \beta_{9+i} D_i \tag{4.7a}$$

Quadratic

$$\begin{aligned} f(X;\beta) = {} & \beta_0 + \beta_1 t + \beta_2 \text{Max}T + \beta_3 \text{Min}T + \beta_4 \text{Rain}T + \beta_5 \text{Max}T^2 + \beta_6 \text{Min}T^2 \\ & + \beta_7 \text{Max}T \times \text{Min}T + \beta_8 \text{Max}T \times \text{Rain}T \\ & + \beta_9 \text{Min}T \times \text{Rain}T + \sum_{i=1}^{S-1} \beta_{9+i} D_i \end{aligned} \tag{4.7b}$$

Variance Function
Exponential:

$$h(X;\delta) = \exp(\delta_0 + \delta_1 t + \delta_2 \text{Max}T + \delta_3 \text{Min}T + \delta_4 \text{Rain}T) \tag{4.8}$$

So Model 2 has two submodels:

(i) Linear mean function and exponential variance function.
(ii) Quadratic mean and exponential variance function.

Model 3

$$f(X;\beta) = \beta_0 + \beta_1 t + \beta_2 T + \beta_3 SDT + \beta_4 P + \beta_5 SDP + \sum_{i=1}^{s} \beta_{5+i}(T \times \text{State}_i)$$

$$+ \sum_{i=1}^{\$1} \beta_{5+S+i}(P \times \text{State}_i)_i + \sum_{i=1}^{S-1} \beta_{5+2s+i} D_i \tag{4.9}$$

Variance Function
Same as the mean function with and without interaction terms.

In our present study, we restrict ourselves with Model 2 wherein we use maximum temperature, minimum temperature and rainfall as climatic variables. As already stated, these variables have been used in recent studies to estimate the impact of climate change on rice yield and the best model among all the submodels (linear or quadratic) is selected based on Akaike Information Criterion (AIC) and other model characteristics.

c. **Akaike Information Criterion (AIC)**

Introduced by Akaike, Hirotugu (1973) provides a means of selecting the best model. This selection is based on the AIC score given by

$$\text{AIC} = 2K - 2\ln(L) \tag{4.10}$$

where K is the number of parameters in the model (including constant and variance) and L is the maximized value of the likelihood function of the model. When we have small samples, that is, when $n/K < 40$, we apply a correction and the corrected AIC is given by

$$\text{AIC}_c = \text{AIC} + \frac{2K(K+1)}{n-K-1} \tag{4.11}$$

In the above model, n is the number of observations. The best model is one for which the AIC or AIC_c is a minimum.

d. **Fixed Effect or Random Effect Model**

Panel data is commonly analysed using either fixed effect or random effect models (Baltagi 2005). In the present study, fixed effects models are purposively used. The main reason is that it helps to estimate state-specific effects; we can include dummy variables for the states in the mean and variance functions. This will accommodate the gap between mean and variance functions of states. Another reason, as pointed out by Barnwal and Kotani (2010), Sarker et al. (2012), is that there may be association between unobserved time invariant properties and the variables included in the study. The presence of such correlations violates the random effects model assumptions. For example, as emphasized by Barnwal and Kotani (2010), states with relatively more suitable climate may have developed better irrigation infrastructure and better soil fertility over time. In our present study, we have selected one district in each state which has best performance in terms of yield during a period of time. Hence, it is more likely that these districts have developed better soil fertility and irrigation sources. Hence, it is reasonable to assume that there is good correlation between the yield and these unobserved time invariant characteristics. So the assumptions of random effects model are violated. So fixed effects models are more suitable, and they will provide unbiased estimates. Studies by McCarl et al. (2008), Cabas et al. (2010), Barnwal and Kotani (2010) and Sarker et al. (2012) also support the choice of fixed effects model, and so the same model is employed in the present study also.

e. **Panel Root Tests**

Just and Pope production function assumes that the variables in the model are stationary (Chen et al. 2004). If this assumption is violated, it will give spurious results (Chen and Chang 2005; Granger and Newbold 1974). Granger and Newbold (1974) showed that deterministic and stochastic trends in the variables can induce spurious correlations between the variables. Time series data with more than 20 years should be tested for stationarity because a spurious result could arise because of non-stationarity on the time series variable in question (Chen et al. 2004; McCarl et al. 2008; Sarkar et al. 2012). Even though inclusion of time trend will minimize this problem, it may not completely solve it when spurious correlations are present. So the variables included in the panel data must be checked for stationarity before we estimate the model using FGLS or MLE method. That is, it is important to test for the presence of unit root for each variable prior to estimation of the model. If the variable is non-stationary, i.e. if they possess I(1) property[1], then they must be differenced before we perform the analysis. Traditional unit root tests (e.g. Dickey–Fuller (DF) test and augmented Dickey–Fuller (ADF) test) are suitable for testing temporal series data on single variable only. They cannot be applied for panel data where we have time series data on different cross sections. But a number of panel root tests have been developed and some of them are developed recently.

The main objective of these tests is to test whether a given series (panel) is stationary for all the individual panel units (in our study, states belonging to the region). In all these tests, the given panel data is modelled by the structure

$$y_{\text{it}} = (1 - \phi_i)\mu_i + \phi_i y_{\text{it}-1} + \varepsilon_{\text{it}}, i = 1, 2, \ldots N; t = 1, 2, \ldots T \tag{4.12}$$

Rearranging the terms, the above equation can be written as

$$\Delta y_{\text{it}} = \alpha_i + \beta_i y_{\text{it}-1} + \varepsilon_{\text{it}}, i = 1, 2, .N; t = 1, 2, \ldots T \tag{4.13}$$

[1]A time series is defined to be stationary, if its statistical properties do not change over time. For example, for a stationary time series mean and variance remain constants. A time series which is not stationary can be converted into a stationary time series by differencing. If a time series $\{y_t\}$ is not stationary, but, if the differenced series $\{y_t - y_{t-1}\}$ is stationary, then the original series $\{y_t\}$ is said to possess $I(1)$ property.

where

$$\beta_i = \phi_i - 1, i = 1, 2, \ldots N \tag{4.14}$$

We want to test the hypothesis that $\phi_i = 1$ for all i. That is, we want to test for unit root. This is equivalent to testing the hypothesis that $H_0 : \beta_i = 0$ for all i.

There are a number of panel unit root tests. Hurlin and Mignon (2007) provide an overview of such tests. The popular tests are Levin, Lin and Chu (LLC) 2002; Im, Pesaran and Shin (IPS) 2003 and Fisher's test proposed by Maddala and Wu (1999) and Choi (2001). Fisher's unit root tests are widely applied. For example, Sarker et al. (2012) have applied this test as it provides more precise results and achieves higher power compared to other tests. Barnwal and Kotani (2010) have used the same test for performing panel root tests. This test combines the p-values of the unit root test statistics in each cross-sectional unit and this test, unlike LLC and IPS tests, does not require the panel data to be balanced. In the present study, the same test has been used for panel unit root testing. The null and alternative hypotheses are:

H_0: All panels contain unit roots
H_a: At least one panel is stationary.

Panel root tests are applied under two models: a) with individual effects without trend (i.e. with intercept only) and b) with individual effects with trend. The second option is suitable when we believe the presence of trend in the concerned variable. For our panel data, we have data for about 25–30 years. Hence, it is important to check for the presence of trend before choosing model. To perform the panel root tests, the MATLAB coding by Hurlin C., and downloaded from the website http://www.execandshare.org/CompanionSite, was used. This MATLAB coding computes the Fisher panel unit root tests, proposed by Choi (2001) and Maddala and Wu (1999). Both tests combine the significant levels obtained from individual augmented Dickey–Fuller (ADF) tests.

f. **Quantile Regression**

One important question is whether climate change factors affect the quantiles of the yield distribution uniformly or differently. Usually, it is hypothesized that lower levels of yield are more likely to be sensitive for changes in climate variables (Barnwal and Kotani (2010). In order to address the above stated hypothesis, this study uses quantile regression. Quantile regressions are variations of ordinary least squares (OLS). The relationship between average yield and a set of explanatory variables is developed in OLS, whereas quantile regression finds the relationship between a quantile yield (e.g. median yield, 25% quartile yield, 75% quartile yield, etc.) and the given set of explanatory variables. It was first introduced by Koenker and Bassett (1978). In our study, it can be used to explain how a given quantile k ($0<k<1$) of the rice yield changes due to changes in one or more climatic variables. It has been shown that when the probability distribution of the response variable (in our study rice yield) is non-normal, quantile regression estimators are more efficient than OLS estimators (Buchinsky 1998). So when the yield data of crops is non-normally distrusted, quantile regression estimators give deeper insights into the relationship between yield and climate variables. The quantile regression model can be specified as

$$y_i = X_i\beta_\theta + u_{6i} \text{ with } \mathrm{Quant}_\theta(y_i|X_i) = X_i\beta_\theta \tag{4.15}$$

where $\mathrm{Quant}_\theta(y_i|X_i)$ represents θth conditional quantile of rice yield y and X denotes a set of explanatory variables, trend, temperature, rainfall and state dummies. The quantile regression functions can be fitted by using Gretl software package.

4.6 Summary

This chapter summarizes the methodologies applied in the present study for quantifying the climate change impacts on rice production. A total of 1707 farmers from 13 states, grouped into 5 regions, were selected for the study. Garrett Ranking technique, Friedman Rank Test and Kendall's coefficient of concordance were used as the statistical tools to analyse primary data collected on farmers' perception of climate change. Econometric tools such as Just–Pope production function, Akaike Information Criterion, fixed effect model and panel root tests are employed in the current study to assess the economic impact of climate change on rice production. A review of past studies on economic impact of climate change on crop yields is also presented in the chapter.

Appendix: Sample Gretl Scripts, Do File and Sample Outputs

1. **Sample Gretl Script to estimate Just–Pope production function Using FGLS**

```
# Estimation of Just Pope Production Function Using FGLS
# Script by Dr.C.R.Ranganathan 18-08-14
#
# Opening the Data File
open G:\CRRDataFiles\NorthRAvgJPData.gdt
# Generating Square, Interaction of CC variables and Dummy Variables for States
genr AvgT2=AvgT^2
genr RainT2=RainT^2
genrAvgTRainT=AvgT*RainT
genr Punjab=(StateID=1)
genr Haryana=(StateID=2)
# First Stage ols and residuals
ols Yield const Time AvgTRainT AvgT2 RainT2 AvgTRainT \
Punjab Haryana
series uh= $uhat
```

```
# Second Statge: Estimation of Variance Function
genr Luh2=log(uh^2)
ols Luh2 const Time AvgTRainT Punjab Haryana
# Third Stage Weighted Least Square Method to estimate mean function
seriesLr=$yhat
genr W=1/exp(Lr)
wls W Yield const Time AvgTRainT AvgT2 RainT2 AvgTRainT \
Punjab Haryana
```

Sample Output

```
gretl version 1.9.91
Current session: 2014-09-30 12:49

# Estimation of Just Pope Production Function Using FGLS
# Script by Dr.C.R.Ranganathan 18-08-14
#
# Opening the Data File
? open G:.gdt

Read datafile G:.gdt
periodicity: 1, maxobs: 93
observations range: 1 to 93

Listing 6 variables:
  0) const      1) StateID    2) Yield      3) Time       4) AvgT
  5) RainT

# Generating Square, Interaction of CC variables and Dummy Variables for States
? genr AvgT2=AvgT^2
Generated series AvgT2 (ID 6)
? genr RainT2=RainT^2
Generated series RainT2 (ID 7)
? genr AvgTRainT=AvgT*RainT
Generated series AvgTRainT (ID 8)
? genr Punjab=(StateID=1)
Generated series Punjab (ID 9)
? genr Haryana=(StateID=2)
Generated series Haryana (ID 10)
# First Stage ols and residuals
? ols Yield const Time AvgT RainT AvgT2 RainT2 AvgTRainT Punjab Haryana

Model 1: OLS, using observations 1-93
Dependent variable: Yield
              coefficient     std. error     t-ratio     p-value
  ---------------------------------------------------------------
  const         0.785161      35.3682         0.02220    0.9823
  Time          0.0380004      0.00416873     9.116      3.47e-014 ***
  AvgT          0.291484       2.38967        0.1220     0.9032
  RainT        -0.00751912     0.00503529    -1.493      0.1391
  AvgT2        -0.00841372     0.0404213     -0.2082     0.8356
  RainT2        1.15559e-06    4.51529e-07    2.559      0.0123    **
  AvgTRainT     0.000195514    0.000171757    1.138      0.2582
  Punjab        1.58892        0.149044      10.66       2.80e-017 ***
  Haryana       0.554109       0.193199       2.868      0.0052    ***

Mean dependent var   2.854064   S.D. dependent var   0.835066
Sum squared resid    9.807372   S.E. of regression   0.341693
R-squared            0.847130   Adjusted R-squared   0.832571
F(8, 84)             58.18564   P-value(F)           4.82e-31
Log-likelihood      -27.36115   Akaike criterion     72.72230
Schwarz criterion    95.51570   Hannan-Quinn         81.92563

Excluding the constant, p-value was highest for variable 4 (AvgT)
```

```
? series uh= $uhat
Generated series uh (ID 11)
# Second Statge: Estimation of Variance Function
? genr Luh2=log(uh^2)
Generated series Luh2 (ID 12)
? ols Luh2 const Time AvgT RainT Punjab Haryana

Model 2: OLS, using observations 1-93
Dependent variable: Luh2

             coefficient   std. error    t-ratio   p-value
  ---------------------------------------------------------
  const      -2.15791      13.3502       -0.1616   0.8720
  Time        0.0163648     0.0268282     0.6100   0.5435
  AvgT       -0.0942311     0.436214     -0.2160   0.8295
  RainT       0.00179283    0.00131444    1.364    0.1761
  Punjab     -0.330535      0.874377     -0.3780   0.7063
  Haryana    -0.328832      1.15343      -0.2851   0.7763

Mean dependent var   -3.604329   S.D. dependent var    2.270875
Sum squared resid     437.1080   S.E. of regression    2.241479
R-squared             0.078672   Adjusted R-squared    0.025722
F(5, 87)              1.485777   P-value(F)            0.202705
Log-likelihood       -203.9238   Akaike criterion      419.8476
Schwarz criterion     435.0432   Hannan-Quinn          425.9831

Excluding the constant, p-value was highest for variable 4 (AvgT)

#ols Luh2 const Time AvgT RainT
# Third Stage Weighted Least Square Method to estimate mean function
? series Lr=$yhat
Generated series Lr (ID 13)
? genr W=1/exp(Lr)
Generated series W (ID 14)
? wls W Yield const Time AvgT RainT AvgT2 RainT2 AvgTRainT Punjab Haryana

Model 3: WLS, using observations 1-93
Dependent variable: Yield
Variable used as weight: W

               coefficient     std. error     t-ratio      p-value
  ------------------------------------------------------------------
  const        -2.37226       26.3066        -0.09018    0.9284
  Time          0.0393390      0.00375067    10.49        6.15e-017 ***
  AvgT          0.458150       1.75891        0.2605     0.7951
  RainT        -0.00402308     0.00547036    -0.7354     0.4641
  AvgT2        -0.0106407      0.0293972     -0.3620     0.7183
  RainT2        8.59025e-07    4.91175e-07    1.749      0.0840    *
  AvgTRainT     9.13326e-05    0.000186397    0.4900     0.6254
  Punjab        1.64280        0.164616       9.980      6.38e-016 ***
  Haryana       0.541451       0.193409       2.800      0.0063    ***
Statistics based on the weighted data:

Sum squared resid     325.9598   S.E. of regression    1.969892
R-squared             0.877397   Adjusted R-squared    0.865720
F(8, 84)              75.14196   P-value(F)            5.05e-35
Log-likelihood       -190.2804   Akaike criterion      398.5608
Schwarz criterion     421.3542   Hannan-Quinn          407.7641

Statistics based on the original data:

Mean dependent var    2.854064   S.D. dependent var    0.835066
Sum squared resid     9.986056   S.E. of regression    0.344792

Excluding the constant, p-value was highest for variable 4 (AvgT)
```

2. Sample Gretl Script to estimate Just–Pope production function using MLE

```
# Estimation of Just Pope Production Function Using FGLS
# Script by Dr.C.R.Ranganathan 21-09-14
#
# Opening the Data File
open G:\CRRDataFiles\NorthRJPData.gdt
# Generating Square, Interaction of CC variables and Dummy Variables for States
genr Punjab=(StateID=1)
genr Haryana=(StateID=2)
genr MaxT2=MaxT^2
genr MinT2=MinT^2
genr RainT2=RainT^2
genr MaxTMinT=MaxT*MinT
genr MaxTRainT=MaxT*RainT
genr MinTRainT=MinT*RainT
# First Stage ols and residuals
ols Yield const Time MaxT MinT RainT  MaxT2 MinT2 RainT2 MaxTMinT \
MaxTRainT MinTRainT Punjab Haryana
series uh= $uhat
# Second Statge: Estimation of Variance Function
genr Luh2=log(uh^2)
ols Luh2 const Time MaxT MinT RainT Punjab Haryana
#genr elastTV=$coeff(AvgT)*mean(AvgT)
#genr elastRV=$coeff(RainT)*mean(RainT)
scalar a0=$coeff(const)
scalar a1=$coeff(Time)
scalar a2=$coeff(MaxT)
scalar a3=$coeff(MinT)
scalar a4=$coeff(RainT)
# Third Stage Weighted Least Square Method to estimate mean function
series Lr=$yhat
genr W=1/exp(Lr)
wls W Yield const Time MaxT MinT RainT  MaxT2 MinT2 RainT2 MaxTMinT \
MaxTRainT MinTRainT Punjab Haryana
#scalar elastTE=$coeff(AvgT)*mean(AvgT)/mean(Yield)
#scalar elastRE=$coeff(RainTcm)*mean(RainTcm)/mean(Yield)
scalar b0=$coeff(const)
scalar b1=$coeff(Time)
scalar b2=$coeff(MaxT)
scalar b3=$coeff(MinT)
scalar b4=$coeff(RainT)
scalar b5=$coeff(MaxT2)
scalar b6=$coeff(MinT2)
scalar b7=$coeff(RainT2)
scalar b8=$coeff(MaxTMinT)
scalar b9=$coeff(MaxTRainT)
scalar b10=$coeff(MinTRainT)
scalar b11=$coeff(Punjab)
scalar b12=$coeff(Haryana)
set bfgs_richardson on
mle logl=-0.5*log(hv) + log(dnorm(Z))
    series mf = b0 + b1*Time +b2*MaxT + b3*MinT + b4*RainT + b5*MaxT2 \
      +b6*MinT2 + b7*RainT2 + b8*MaxTMinT + b9*MaxTRainT +b10*MinTRainT \
      + b11*Punjab + b12*Haryana
    series hv =exp(a0 +a1*Time +a2*MaxT + a3*MinT +a4*RainT)
    series Z =(Yield-mf)/sqrt(hv)
    deriv a0 =-0.5*(1-Z^2)
```

```
      deriv a1 =-0.5*(1-Z^2)*Time
      deriv a2 =-0.5*(1-Z^2)*MaxT
      deriv a3 =-0.5*(1-Z^2)*MinT
      deriv a4 =-0.5*(1-Z^2)*RainT
      deriv b0 =Z/sqrt(hv)
      deriv b1 =(Z/sqrt(hv))*Time
      deriv b2 =(Z/sqrt(hv))*MaxT
      deriv b3 =(Z/sqrt(hv))*MinT
      deriv b4 =(Z/sqrt(hv))*RainT
      deriv b5 =(Z/sqrt(hv))*MaxT2
      deriv b6 =(Z/sqrt(hv))*MinT2
      deriv b7 =(Z/sqrt(hv))*RainT2
      deriv b8 =(Z/sqrt(hv))*MaxTMinT
      deriv b9 =(Z/sqrt(hv))*MaxTRainT
      deriv b10 =(Z/sqrt(hv))*MinTRainT
      deriv b11 =(Z/sqrt(hv))*Punjab
      deriv b12 =(Z/sqrt(hv))*Haryana
  end mle
```

Sample Output

```
gretl version 1.9.91
Current session: 2014-09-30 12:51

# Estimation of Just Pope Production Function Using FGLS
# Script by Dr.C.R.Ranganathan 21-09-14
#
# Opening the Data File
? open G:.gdt

Read datafile G:.gdt
periodicity: 31, maxobs: 93
observations range: 1:01 to 3:31

Listing 9 variables:
  0) const       1) StateID     2) Year       3) Yield       4) Time
  5) AvgT        6) MaxT        7) MinT       8) RainT

# Generating Square, Interaction of CC variables and Dummy Variables for States
? genr Punjab=(StateID=1)
Generated series Punjab (ID 9)
? genr Haryana=(StateID=2)
Generated series Haryana (ID 10)
? genr MaxT2=MaxT^2
Generated series MaxT2 (ID 11)
? genr MinT2=MinT^2
Generated series MinT2 (ID 12)
? genr RainT2=RainT^2
Generated series RainT2 (ID 13)
? genr MaxTMinT=MaxT*MinT
Generated series MaxTMinT (ID 14)
? genr MaxTRainT=MaxT*RainT
Generated series MaxTRainT (ID 15)
? genr MinTRainT=MinT*RainT
Generated series MinTRainT (ID 16)
# First Stage ols and residuals
? ols Yield const Time MaxT MinT RainT MaxT2 MinT2 RainT2 MaxTMinT MaxTRainT \par
MinTRainT Punjab Haryana

Model 1: Pooled OLS, using 93 observations
Included 3 cross-sectional units
Time-series length = 31
Dependent variable: Yield
```

```
             coefficient     std. error     t-ratio      p-value
  ------------------------------------------------------------------
  const        -2.69786      39.3139        -0.06862     0.9455
  Time          0.0424002     0.00409913    10.34        2.08e-016 ***
  MaxT          4.06101       2.46250        1.649       0.1030
  MinT         -5.54125       1.88343       -2.942       0.0043    ***
  RainT         0.00710528    0.00592957     1.198       0.2343
  MaxT2        -0.140689      0.0525708     -2.676       0.0090    ***
  MinT2        -0.0885605     0.0570486     -1.552       0.1245
  RainT2       -3.34060e-07   5.74530e-07   -0.5814      0.5626
  MaxTMinT      0.260415      0.0899561      2.895       0.0049    ***
  MaxTRainT    -0.000970450   0.000258727   -3.751       0.0003    ***
  MinTRainT     0.00109967    0.000291935    3.767       0.0003    ***
  Punjab        1.53502       0.179225       8.565       6.30e-013 ***
  Haryana       0.541471      0.194454       2.785       0.0067    ***

Mean dependent var   2.854064   S.D. dependent var   0.835066
Sum squared resid    7.501095   S.E. of regression   0.306209
R-squared            0.883078   Adjusted R-squared   0.865540
F(12, 80)            50.35151   P-value(F)           3.47e-32
Log-likelihood      -14.89518   Akaike criterion     55.79037
Schwarz criterion    88.71416   Hannan-Quinn         69.08406
rho                  0.363012   Durbin-Watson        1.195375

Excluding the constant, p-value was highest for variable 13 (RainT2)

? series uh= $uhat
Generated series uh (ID 17)
# Second Statge: Estimation of Variance Function
? genr Luh2=log(uh^2)
Generated series Luh2 (ID 18)
? ols Luh2 const Time MaxT MinT RainT Punjab Haryana

Model 2: Pooled OLS, using 93 observations
Included 3 cross-sectional units
Time-series length = 31
Dependent variable: Luh2

             coefficient    std. error    t-ratio     p-value
  --------------------------------------------------------------
  const       -2.17281      12.5729       -0.1728     0.8632
  Time         0.00490119    0.0255307     0.1920     0.8482
  MaxT         0.0260893     0.407787      0.06398    0.9491
  MinT        -0.0810797     0.441439     -0.1837     0.8547
  RainT        0.000257499   0.00129560    0.1987     0.8429
  Punjab      -1.92298       0.954147     -2.015      0.0470   **
  Haryana     -0.499248      1.09491      -0.4560     0.6496

Mean dependent var  -3.792231   S.D. dependent var   2.164226
Sum squared resid    366.0300   S.E. of regression   2.063047
R-squared            0.150578   Adjusted R-squared   0.091316
F(6, 86)             2.540886   P-value(F)           0.025916
Log-likelihood      -195.6717   Akaike criterion     405.3433
Schwarz criterion    423.0715   Hannan-Quinn         412.5015
rho                  0.054128   Durbin-Watson        1.814540

Excluding the constant, p-value was highest for variable 6 (MaxT)

#genr elastTV=$coeff(AvgT)*mean(AvgT)
#genr elastRV=$coeff(RainT)*mean(RainT)
? scalar a0=$coeff(const)
Generated scalar a0 = -2.17281
? scalar a1=$coeff(Time)
Generated scalar a1 = 0.00490119
```

```
? scalar a2=$coeff(MaxT)
Generated scalar a2 = 0.0260893
? scalar a3=$coeff(MinT)
Generated scalar a3 = -0.0810797
? scalar a4=$coeff(RainT)
Generated scalar a4 = 0.000257499
# Third Stage Weighted Least Square Method to estimate mean function
? series Lr=$yhat
Generated series Lr (ID 19)
? genr W=1/exp(Lr)
Generated series W (ID 20)
? wls W Yield const Time MaxT MinT RainT MaxT2 MinT2 RainT2 MaxTMinT MaxTRainT \par
MinTRainT Punjab Haryana

Model 3: WLS, using 93 observations
Dependent variable: Yield
Variable used as weight: W

              coefficient     std. error    t-ratio    p-value
  -------------------------------------------------------------
  const        27.0829        37.3277         0.7255   0.4702
  Time          0.0488063      0.00321416    15.18     2.81e-025 ***
  MaxT          2.26438        2.42206        0.9349   0.3527
  MinT         -5.46233        1.86166       -2.934    0.0044    ***
  RainT         0.00542750     0.00583105     0.9308   0.3548
  MaxT2        -0.116323       0.0543280     -2.141    0.0353    **
  MinT2        -0.100708       0.0608617     -1.655    0.1019
  RainT2       -4.67472e-07    4.20998e-07   -1.110    0.2702
  MaxTMinT      0.270352       0.0995194      2.717    0.0081    ***
  MaxTRainT    -0.00102960     0.000248053   -4.151    8.22e-05  ***
  MinTRainT     0.00126102     0.000297232    4.243    5.90e-05  ***
  Punjab        1.43799        0.204759       7.023    6.41e-010 ***
  Haryana       0.469765       0.230385       2.039    0.0447    **

Statistics based on the weighted data:

Sum squared resid     287.1216   S.E. of regression   1.894471
R-squared             0.920553   Adjusted R-squared   0.908636
F(12, 80)             77.24636   P-value(F)           8.22e-39
Log-likelihood       -184.3810   Akaike criterion     394.7621
Schwarz criterion     427.6859   Hannan-Quinn         408.0557

Statistics based on the original data:

Mean dependent var    2.854064   S.D. dependent var   0.835066
Sum squared resid     7.809091   S.E. of regression   0.312432

Excluding the constant, p-value was highest for variable 8 (RainT)

#scalar elastTE=$coeff(AvgT)*mean(AvgT)/mean(Yield)
#scalar elastRE=$coeff(RainTcm)*mean(RainTcm)/mean(Yield)
? scalar b0=$coeff(const)
Generated scalar b0 = 27.0829
? scalar b1=$coeff(Time)
Generated scalar b1 = 0.0488063
? scalar b2=$coeff(MaxT)
Generated scalar b2 = 2.26438
? scalar b3=$coeff(MinT)
Generated scalar b3 = -5.46233
? scalar b4=$coeff(RainT)
Generated scalar b4 = 0.0054275
? scalar b5=$coeff(MaxT2)
Generated scalar b5 = -0.116323
? scalar b6=$coeff(MinT2)
Generated scalar b6 = -0.100708
? scalar b7=$coeff(RainT2)
Generated scalar b7 = -4.67472e-007
? scalar b8=$coeff(MaxTMinT)
```

```
Generated scalar b8 = 0.270352
? scalar b9=$coeff(MaxTRainT)
Generated scalar b9 = -0.0010296
? scalar b10=$coeff(MinTRainT)
Generated scalar b10 = 0.00126102
? scalar b11=$coeff(Punjab)
Generated scalar b11 = 1.43799
? scalar b12=$coeff(Haryana)
Generated scalar b12 = 0.469765
? set bfgs_richardson on
? mle logl=-0.5*log(hv) + log(dnorm(Z))
? series mf = b0 + b1*Time +b2*MaxT + b3*MinT + b4*RainT + b5*MaxT2 +b6*MinT2 \par
+ b7*RainT2 + b8*MaxTMinT + b9*MaxTRainT +b10*MinTRainT + b11*Punjab + \par
b12*Haryana
? series hv =exp(a0 +a1*Time +a2*MaxT + a3*MinT +a4*RainT)
? series Z =(Yield-mf)/sqrt(hv)
? deriv a0 =-0.5*(1-Z^2)
? deriv a1 =-0.5*(1-Z^2)*Time
? deriv a2 =-0.5*(1-Z^2)*MaxT
? deriv a3 =-0.5*(1-Z^2)*MinT
? deriv a4 =-0.5*(1-Z^2)*RainT
? deriv b0 =Z/sqrt(hv)
? deriv b1 =(Z/sqrt(hv))*Time
? deriv b2 =(Z/sqrt(hv))*MaxT
? deriv b3 =(Z/sqrt(hv))*MinT
? deriv b4 =(Z/sqrt(hv))*RainT
? deriv b5 =(Z/sqrt(hv))*MaxT2
? deriv b6 =(Z/sqrt(hv))*MinT2
? deriv b7 =(Z/sqrt(hv))*RainT2
? deriv b8 =(Z/sqrt(hv))*MaxTMinT
? deriv b9 =(Z/sqrt(hv))*MaxTRainT
? deriv b10 =(Z/sqrt(hv))*MinTRainT
? deriv b11 =(Z/sqrt(hv))*Punjab
? deriv b12 =(Z/sqrt(hv))*Haryana
? end mle
Using analytical derivatives
Tolerance = 1.81899e-012

Function evaluations: 332
Evaluations of gradient: 74

Model 4: ML, using 93 observations
logl = -0.5*log(hv) + log(dnorm(Z))
Standard errors based on Outer Products matrix
```

	estimate	std. error	z	p-value	
a0	4.95313	8.19414	0.6045	0.5455	
a1	0.00893487	0.0260521	0.3430	0.7316	
a2	-0.525693	0.358395	-1.467	0.1424	
a3	0.407885	0.443294	0.9201	0.3575	
a4	0.000351394	0.000951061	0.3695	0.7118	
b0	-7.49489	39.7426	-0.1886	0.8504	
b1	0.0455863	0.00384772	11.85	2.21e-032	***
b2	4.20032	2.68857	1.562	0.1182	
b3	-5.40478	2.05361	-2.632	0.0085	***
b4	0.00994506	0.00641027	1.551	0.1208	
b5	-0.151541	0.0667235	-2.271	0.0231	**
b6	-0.110422	0.0687613	-1.606	0.1083	
b7	-4.62258e-07	5.08444e-07	-0.9092	0.3633	
b8	0.287166	0.115094	2.495	0.0126	**
b9	-0.00104226	0.000294570	-3.538	0.0004	***
b10	0.00109221	0.000317638	3.439	0.0006	***
b11	1.50951	0.148421	10.17	2.69e-024	***
b12	0.508953	0.160703	3.167	0.0015	***

```
Log-likelihood      -9.743832   Akaike criterion     55.48766
Schwarz criterion    101.0745   Hannan-Quinn         73.89431
```

3. Sample Do file and executing it for obtaining MLEs

```
. use
"G:\FDriveFiles\13StatesCCStudy\DataRegionwiseAnalysed\NorthR\NorthRJPData.dta"

. do "G:\FDriveFiles\RIHNFellowship\JustANDPope-Model-Analysis\myjpmle1.do"

. program define myjpmle1
  1.         version 1.0
  2.         args lnf Xb Xd
  3.         quietly replace `lnf' = -`Xd'/2+ln(normd(($ML_y1-
`Xb')/sqrt(exp(`Xd'))))
  4. end

.
end of do-file

. ml model lf myjpmle1 ( yield =  time maxt mint raint maxt2 mint2 raint2 maxtmint
maxtraint mintraint punjab harya
> na) ( yield =  time maxt mint raint)

. ml maximize

initial:        log likelihood = -496.43998
alternative:    log likelihood = -284.52857
rescale:        log likelihood = -187.40136
rescale eq:     log likelihood = -148.53285
Iteration 0:    log likelihood = -148.53285  (not concave)
Iteration 1:    log likelihood = -92.928659  (not concave)
Iteration 2:    log likelihood = -63.238579  (not concave)
Iteration 3:    log likelihood = -50.906082  (not concave)
Iteration 4:    log likelihood = -44.778912  (not concave)
Iteration 5:    log likelihood = -33.377034
Iteration 6:    log likelihood =  -20.03102
Iteration 7:    log likelihood = -11.306557
Iteration 8:    log likelihood = -9.7577626
Iteration 9:    log likelihood = -9.7477721
Iteration 10:   log likelihood = -9.7477702
Iteration 11:   log likelihood = -9.7477702

                                                  Number of obs   =         93
                                                  Wald chi2(12)   =     817.32
Log likelihood = -9.7477702                       Prob > chi2     =     0.0000

------------------------------------------------------------------------------
             |      Coef.   Std. Err.      z    P>|z|     [95% Conf. Interval]
-------------+----------------------------------------------------------------
eq1          |
        time |    .045632   .0036608    12.47   0.000      .038457     .052807
        maxt |   4.217792   2.395352     1.76   0.078    -.4770114    8.912596
        mint |  -5.373193   1.734861    -3.10   0.002    -8.773458   -1.972929
       raint |   .0098913   .0056688     1.74   0.081    -.0012193    .0210019
       maxt2 |  -.1519693   .0540734    -2.81   0.005    -.2579512   -.0459873
       mint2 |  -.1113685   .0576984    -1.93   0.054    -.2244554    .0017184
      raint2 |  -4.61e-07   5.06e-07    -0.91   0.362    -1.45e-06    5.31e-07
    maxtmint |    .287581   .0932881     3.08   0.002     .1047398    .4704223
   maxtraint |  -.0010415   .0002469    -4.22   0.000    -.0015255   -.0005576
   mintraint |   .0010933   .0002731     4.00   0.000     .0005581    .0016285
      punjab |   1.511339   .1582188     9.55   0.000     1.201235    1.821442
     haryana |   .5099374   .1773931     2.87   0.004     .1622534    .8576214
       _cons |  -8.144124   35.03346    -0.23   0.816    -76.80845     60.5202
-------------+----------------------------------------------------------------
eq2          |
        time |   .0089379   .0209578     0.43   0.670    -.0321386    .0500144
        maxt |  -.5244976   .2692086    -1.95   0.051    -1.052137    .0031416
        mint |    .400668   .3068341     1.31   0.192    -.2007157    1.002052
       raint |   .0003664   .0008137     0.45   0.653    -.0012285    .0019612
       _cons |   5.076084   5.472121     0.93   0.354    -5.649077    15.80124
------------------------------------------------------------------------------
```

```
. estat ic

-----------------------------------------------------------------------------
       Model |        Obs  ll(null)  ll(model)      df         AIC        BIC
-------------+---------------------------------------------------------------
           . |         93         .  -9.74777       18    55.49554   101.0823
-----------------------------------------------------------------------------
               Note:  N=Obs used in calculating BIC; see [R] BIC note
```

Chapter 5
Climate Change Impacts on Rice Yield in Northern Region of India

K. Palanisami, Krishna Reddy Kakumanu, Udaya Sekhar Nagothu and C. R. Ranganathan

5.1 Introduction

The north region for the present study covers three states: Punjab, Haryana and Uttar Pradesh. The districts selected from the three states were: Sangrur, Karnal and Chandauli, respectively. As stated in the methodology chapter, the present study is based on (i) primary data: farm survey among 100 farmers from each of the three districts and (ii) secondary data: time series data (1980–2010) on productivity (yield) of rice and climate variables from the three districts. The location of the study area is given in Fig. 5.1.

5.2 Socio-economic Profiles of the Farmers

All the 300 farmers surveyed were males. Out of them, 200 farmers in Punjab and Haryana have crop with supplemental irrigation typology, while all the 100 farmers in Uttar Pradesh have irrigated crop-based farming. The average ages of the farmers in the three states are 39, 40 and 53 years, respectively. So, farmers in Uttar Pradesh are relatively older than those from Punjab and Haryana. About 23% of the farmers have no formal education, 21% have studied up to primary level, and the rest 56% have studied up to secondary level (Fig. 5.2); 86% of the farmers from Uttar Pradesh have education up to secondary level.

5.3 Farm Information

Farmers in the study region own land and wells. They use different methods of irrigation like drip, sprinkler and others, and they invest on digging borewells. They cultivate rice crop only during the kharif season. Farmers in Punjab and Haryana

K. Palanisami et al., *Climate Change and Future Rice Production in India*, India Studies in Business and Economics,
https://doi.org/10.1007/978-981-13-8363-2_5

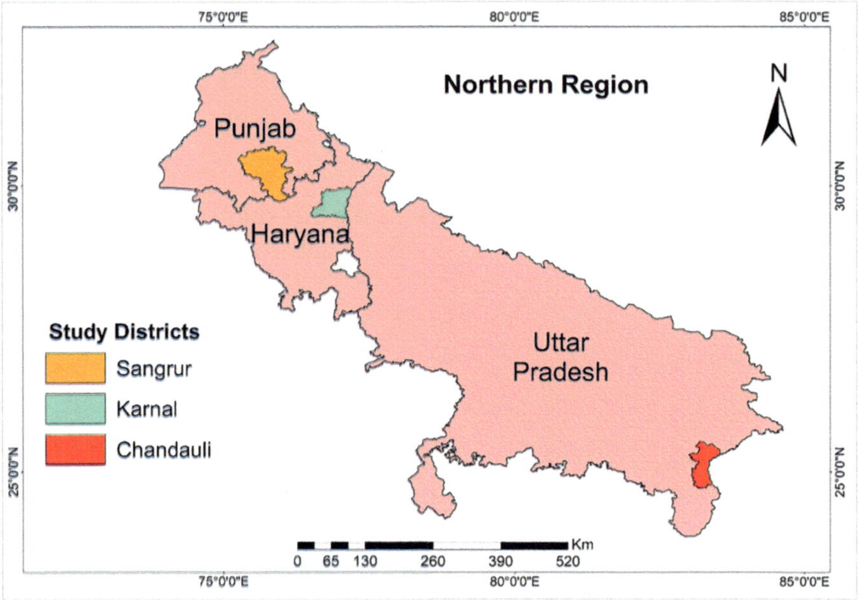

Fig. 5.1 Study regions

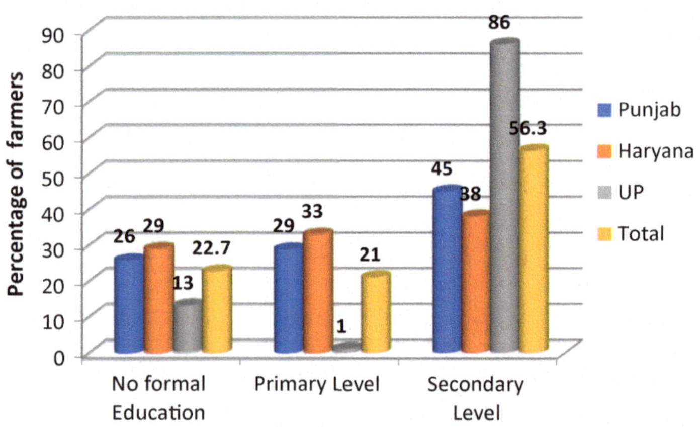

Fig. 5.2 Educational status of the sample farmers

Table 5.1 Summary statistics of irrigated land owned (hectares)

State	Mean	S.D.	Min	Max
Punjab	4.98	6.35	0.40	60.70
Haryana	4.01	2.79	0.61	16.19
UP	2.10	1.74	0.40	8.09
Overall	3.68	4.29	0.40	60.70

S.D standard deviation

Table 5.2 Distribution of number of borewells owned across states

Number of farmers owning				
No. of borewells	Punjab	Haryana	UP	Total
0	0	0	1	1
1	59	73	92	224
2	34	25	7	66
3	6	2	0	8
> 3	1	0	0	1

derive income from rice and other crops only during kharif season as they are not raising any crop during rabi and summer. Though farmers in Uttar Pradesh could get income from rice crop only in kharif, they could get income from other crops also during all the seasons.

(a) **Physical assets**:

(i) **Land owned**: All the farmers own only irrigated land. Table 5.1 gives the summary statistics.

The mean area under irrigation varies from 2.10 to 4.98 ha across the three states, and the overall mean is 3.68 ha. Landholdings in Punjab have the largest standard deviation of 6.35 ha, and the minimum and maximum holdings are, respectively, 0.40 and 60.7 ha.

(ii) **Wells owned**: No farmer possesses open well, and they own only borewells. Table 5.2 gives the frequency distribution of number of borewells owned by the farmers.

Out of 300 farmers, 224 of them, that is, 75% of the farmers, possess one borewell, and in Uttar Pradesh, 92% of the farmers possess on borewell each. Also, overall, 22% of the farmers possess 2 borewells each.

(iii) **Sources of irrigation**: Farmers have two sources of irrigation, viz. canal and well.

All farmers of Haryana have well and that is the main source of their irrigation, whereas in Punjab, 61 of them use wells and the remaining 39 use both canal and well. But in Uttar Pradesh, 86% of the farmers use both sources and only 13% use well (Fig. 5.3).

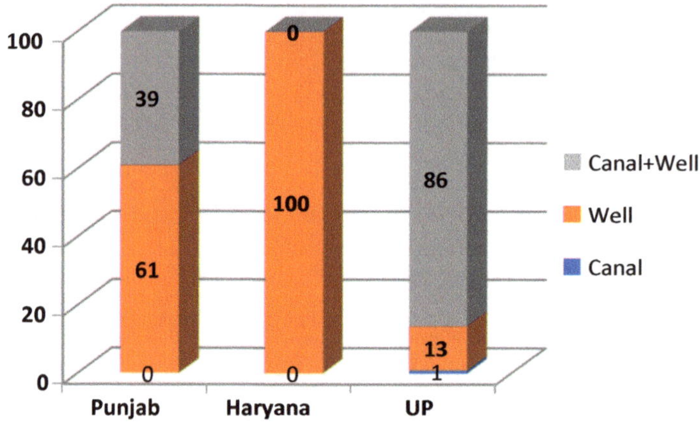

Fig. 5.3 Frequency distribution of sources of irrigation

(iv) **Investment in irrigation**: None of the farmers in Haryana made any investment on irrigation structures and farmers in Punjab and Uttar Pradesh invested in digging borewells. On the average, farmers in Punjab invested Rs. 34,800 and farmers in Uttar Pradesh spent, on the average, Rs. 61,630 in digging borewells.

(v) **Area under rice**: All the farmers have cultivated rice only during the kharif season. During this season, on the average every farmer cultivates rice in 4.43, 3.84 and 2.02 ha in the three states, respectively. Similarly, other crops are grown, respectively, in 0.071, 0.41 and 0.53 ha in the three states during the same season. No crop is cultivated by the sample farmers during rabi and summer season in Punjab and Haryana, whereas in Uttar Pradesh, on the average, other crops are grown in 2.1 and 1.34 ha, respectively, during these two seasons.

5.4 Climate Change Events, Shocks and Their Effects as Perceived by Farmers

With a view to have an in-depth understanding of farmers' perception on the climate change, its effects on agriculture and various adaptation strategies/techniques followed by the farmers, detailed survey was conducted and the results are discussed below.

5.4.1 *Perception of Farmers on Climate Change Events*

Farmers were asked to rank the 16 events as listed in the methodology chapter according to their perceptions of climate change (CC). To find what the farmers perceive as the most significant effects of climate change, Garrett ranking technique was applied. Farmers in Punjab and Uttar Pradesh ranked all the 16 CC events, while farmers in Haryana ranked only 8 CC events. Hence, the rankings of these two states, viz. Punjab and Uttar Pradesh, were combined to represent northern region. Garrett's ranking technique, as described in the methodology chapter, was used. Tables 5.3 and 5.4 give the results of the Garrett Ranking. Table 5.4 shows that untimely rain (UR) is the most important climate change event (Rank 1) perceived by rice farmers in northern region. It earned an average Garrett Score of 66.385 as given in Table 5.4. It was given Ranks 1, 2 and 3 by 31, 41 and 47 farmers, respectively. The second and third important climate change events are, respectively, incidence of pests and diseases (IPD) and depletion of groundwater (DGW).

The results for Haryana are presented in Tables 5.5 and 5.6. Depletion in groundwater is the foremost concern for Haryana farmers as all of them have given Rank 1 to it. The same CC event has been given Rank 3 by farmers of Punjab and Uttar Pradesh. Incidence of pest and disease and untimely rains were given Ranks 2 and 3 by Haryana farmers. The same two CC events were given Ranks 2 and 1 by the farmers of Punjab and Uttar Pradesh. Thus, as a whole, these three events are the most important CC events as perceived by the farmers belonging to the northern region. Similar Garrett ranking analysis was done separately for Punjab and Uttar Pradesh. The results are presented in Appendix (Tables 5.20, 5.21, 5.22, 5.23).

Friedman Rank Test was applied to test for differences between rankings of the climate change events (a listed in Table 5.3) by the farmers. The methodology as described in Chap. 4 was followed. The calculated value of Friedman's F statistics was 672.9, while the table value (corresponding to 15 degrees of freedom at 5% level of significance) is 24.996. This shows that the farmers did not consistently rank the events. Thus, the farmers' perceptions of climate change effects differ across the sample chosen. The same conclusion was arrived by applying Kendall's coefficient of concordance also. Similar analysis was done for individual states, viz. Punjab, Haryana and Uttar Pradesh. The results show that there is no consistency in the rankings of the events by the farmers.

5.4.2 *Effects of Shocks*

Farmers were questioned on their perception about the effects of climatic shocks. They were asked to Rank 7 effects of climatic shocks, as described in methodology chapter. Farmers ranked decline in crop yield (DCY) as the most important shock (Tables 5.7 and 5.8). Out of 300 farmers, 199 of them have given Rank 1 to this effect. The second and third ranks are given to loss of income (LI) and loss of assets

Table 5.3 Factors influencing farmers perception of climate change—Garrett ranking: northern region

Climate change event	Rank																Total no. of farmers
	1	2	3	4	5	6	7	8	9	10	11	12	13	14	15	16	
Drought	2	1	2	6	11	12	9	10	13	9	7	11	13	25	49	20	200
Hailstorm	48	17	11	6	3	4	5	6	3	7	8	12	13	21	22	14	200
Flood	9	36	14	9	1	4	8	5	9	4	7	5	17	16	23	33	200
AD	0	5	5	6	7	6	15	20	23	18	20	14	25	15	13	8	200
IPD	25	32	17	34	15	24	22	11	7	1	2	1	4	2	1	2	200
IFM	3	6	7	3	13	11	13	4	3	6	8	11	5	10	16	81	200
UR	31	41	47	10	14	13	13	6	1	5	3	6	3	4	2	1	200
IW	17	23	26	30	13	16	14	8	9	7	4	4	4	10	10	5	200
TFH	5	15	12	17	15	15	7	14	11	15	24	21	19	5	4	1	200
TFL	1	8	19	13	19	9	6	10	7	13	11	30	30	16	7	1	200
SSDI	0	1	4	7	8	12	28	33	41	21	18	10	7	6	3	1	200
CCP	0	0	1	2	6	8	12	14	11	18	26	30	16	19	24	13	200
CLA	0	0	0	1	4	9	14	19	32	27	21	16	17	30	7	3	200
CFI	0	1	2	24	20	27	14	18	16	24	11	16	14	4	7	2	200
CF	4	5	24	19	33	17	9	8	9	11	24	7	10	4	7	9	200
DGW	55	9	9	13	18	13	11	14	5	14	6	6	3	13	5	6	200
Total	200	200	200	200	200	200	200	200	200	200	200	200	200	200	200	200	3200
Garrett score	86	76	70	65	61	58	55	52	48	45	42	39	35	30	24	14	

Table 5.4 Rankings of factors representing climate change using Garrett ranking: northern region

Climate change event (factors)	Rank																Total	Average	Rank
	1	2	3	4	5	6	7	8	9	10	11	12	13	14	15	16			
	Total score																		
Drought	172	76	140	390	671	696	495	520	624	405	294	429	455	750	1176	280	7573	37.865	15
Hailstorm	4128	1292	770	390	183	232	275	312	144	315	336	468	455	630	528	196	10654	53.27	5
Flood	774	2736	980	585	61	232	440	260	432	180	294	195	595	480	552	462	9258	46.29	11
AD	0	380	350	390	427	348	825	1040	1104	810	840	546	875	450	312	112	8809	44.045	12
IPD	2150	2432	1190	2210	915	1392	1210	572	336	45	84	39	140	60	24	28	12827	64.135	2
IFM	258	456	490	195	793	638	715	208	144	270	336	429	175	300	384	1134	6925	34.625	16
UR	2666	3116	3290	650	854	754	715	312	48	225	126	234	105	120	48	14	13277	66.385	1
IW	1462	1748	1820	1950	793	928	770	416	432	315	168	156	140	300	240	70	11708	58.54	4
TFH	430	1140	840	1105	915	870	385	728	528	675	1008	819	665	150	96	14	10368	51.84	7
TFL	86	608	1330	845	1159	522	330	520	336	585	462	1170	1050	480	168	14	9665	48.325	10
SSDI	0	76	280	455	488	696	1540	1716	1968	945	756	390	245	180	72	14	9821	49.105	9
CCP	0	0	70	130	366	464	660	728	528	810	1092	1170	560	570	576	182	7906	39.53	14
CLA	0	0	0	65	244	522	770	988	1536	1215	882	624	595	900	168	42	8551	42.755	13
CFI	0	76	140	1560	1220	1566	770	936	768	1080	462	624	490	120	168	28	10008	50.04	8
CF	344	380	1680	1235	2013	986	495	416	432	495	1008	273	350	120	168	126	10521	52.605	6
DGW	4730	684	630	845	1098	754	605	728	240	630	252	234	105	390	120	84	12129	60.645	3

Table 5.5 Factors influencing farmer's perception of climate change—Garrett ranking: Haryana

Climate change event	Rank								Total no. of farmers
	1	2	3	4	5	6	7	8	
AD	0	0	0	0	0	67	33	0	100
IPD	0	90	0	0	0	8	2	0	100
IFM	0	8	0	0	0	0	65	27	100
UR	0	2	90	0	8	0	0	0	100
IW	0	0	2	78	0	0	0	20	100
TFH	0	0	8	0	92	0	0	0	100
TFL	0	0	0	22	0	25	0	53	100
DGW	100	0	0	0	0	0	0	0	100
Total	100	100	100	100	100	100	100	100	800
Garrett score	80	67	60	53	47	40	33	20	

Table 5.6 Rankings of factors representing climate change using Garrett ranking: Haryana

Climate change event	Rank										Rank
	1	2	3	4	5	6	7	8	Total score	Average score	
	Total score										
AD	0	0	0	0	0	2680	1089	0	3769	37.69	6
IPD	0	6030	0	0	0	320	66	0	6416	64.16	2
IFM	0	536	0	0	0	0	2145	540	3221	32.21	8
UR	0	134	5400	0	376	0	0	0	5910	59.1	3
IW	0	0	120	4134	0	0	0	400	4654	46.54	5
TFH	0	0	480	0	4324	0	0	0	4804	48.04	4
TFL	0	0	0	1166	0	1000	0	1060	3226	32.26	7
DGW	8000	0	0	0	0	0	0	0	8000	80	1

(LA), respectively. This is understandable because decline in crop yield will reflect in income and asset generation.

5.4.3 Mitigation Strategies Used by Farmers

Farmers apply several mitigation strategies to tide over the negative impacts of climate change. Table 5.9 provides a summary. It shows that 100% of the farmers in all the three states have restricted rice-growing season to one. Farmers in Uttar Pradesh have used only this strategy. All the farmers in Punjab and Haryana have changed crop variety and planting dates. Also 52% of the farmers in Punjab have sold livestock. All other strategies are followed by less than 6% of the farmers.

Table 5.7 Rankings of effects of climatic shocks by farmers

Rank								
Factors	1	2	3	4	5	6	7	Total
DCY	199	26	14	12	8	20	21	300
LA	7	41	81	30	36	81	24	300
LI	61	136	75	8	5	10	5	300
FIS	5	40	21	74	77	43	40	300
DL	5	16	57	100	68	20	34	300
DC	1	3	25	49	47	45	130	300
DH	22	38	27	27	59	81	46	300
Total	300	300	300	300	300	300	300	2100
Garrett score	79	66	57	50	43	34	21	

Table 5.8 Garrett ranking of effects of climatic shocks

Factors	Rank									
	1	2	3	4	5	6	7	Total score	Average score	Rank
	Total score									
DCY	15721	1716	798	600	344	680	441	20,300	67.67	1
LA	553	2706	4617	1500	1548	2754	504	14,182	47.27	3
LI	4819	8976	4275	400	215	340	105	19,130	63.77	2
FIS	395	2640	1197	3700	3311	1462	840	13,545	45.15	5
DL	395	1056	3249	5000	2924	680	714	14,018	46.73	4
DC	79	198	1425	2450	2021	1530	2730	10,433	34.78	7
DH	1738	2508	1539	1350	2537	2754	966	13,392	44.64	6

5.4.4 Opinions of Farmers on Government Initiatives

From time to time, both Central and State governments implement several mitigation and long-term measures to offset the negative effects of climate change on agriculture. With a view to know the awareness of these measures among the farmers, they were questioned on seven government initiatives (as described in the methodology chapter). They were asked to state their replies as either 'yes', or 'no' or 'not aware'. Figure 5.4 gives a pictorial representation of the awareness on these initiatives by the farmers belonging to the entire region. About 93% of the farmers are well aware of introduction of new crop varieties, 60% have good understanding of water regulation by government through irrigation department, and 53% of the farmers have good understanding on the subsidised inputs (like fertilizers, drip and machinery). Only 29% of the farmers participated in the field demos of the listed adaptation practices. Further, weather-based cropping insurance schemes are not popular among the farm-

Table 5.9 Frequency distribution of mitigation strategies followed by farmers

Strategies	Punjab	Haryana	UP	Total no. of farmers
Did nothing	0	0	0	0
Changed crop variety	100	100	0	200
Changed planting date	100	100	0	200
Left land fallow	1	0	0	1
Sold part of land	0	0	0	0
Leased out part of land	5	0	0	5
Restricted number of seasons for rice-growing to 1	100	100	100	300
Sold livestock	52	0	0	52
Shifted to other crops	1	0	0	1
Borrowed money from relatives/others	4	0	0	4
Received assistance from government/NGOs	5	0	0	5
Less food consumption	0	0	0	0
Shifted to non-farm employment	0	0	2	2
Outmigration to cities	0	0	2	2

ers as 59% of them are not aware of it. Similarly, only 5% of the farmers know about establishment of village knowledge centres. Thus, we conclude that except (i) water regulation through irrigation department, (ii) introduction of new crop varieties and (iii) subsidised inputs, all other government initiatives are not much popular among farmers. This conclusion calls for formulating suitable popularisation mechanisms by the government. Table 5.10 provides a summary of the opinions on government initiatives, state-wise.

Chi-square test was applied to check if the opinions differ across the states and the results show that the opinions of the farmers on government initiatives differ strongly between the three states. This may be because the government initiatives depend on local natural resources availability and other factors.

5.5 Analysis of the Effect of Climate Change on Rice Yield

5.5.1 Data

The objective of the study is to estimate the effects of climate variables on the yield of rice crop using historical data. Hence, we used a panel data set of yield and climate variables pertaining to 31 years from 1980 to 2010. The data on temperature corresponds to the growing season of rice variety in the respective states, and the time series data on rainfall was collected for the entire year to account for rainfall

Table 5.10 Farmers opinions on government initiatives

Govt. initiatives	Punjab			Haryana			UP			Total			Chi-square	df	*p*-value
	Yes	No	NA	Yes	No	NA	Yes	No	NA	Yes	No	NA			
Water regulation through irrigation department	81	19	0	100	0	0	0	31	69	181	50	69	260.8	4	<0.00001
Introduction of new crop varieties	79	20	1	100	0	0	99	1	0	278	21	1	41.3	4	<0.00001
Field demos on the listed adaptation practices	37	52	11	27	21	52	24	72	4	88	145	67	90.7	4	<0.00001
Organised field days on the adaptation practices	31	47	22	3	46	51	10	83	7	44	176	80	81.6	4	<0.00001
Subsidised inputs (fertilizer, drip and machinery)	10	48	42	51	41	8	100	0	0	161	89	50	180.7	4	<0.00001
Weather-based crop insurance schemes	3	43	54	46	33	21	0	0	100	49	76	175	175.1	4	<0.00001
Establishment of village knowledge centres, climate cell, etc.	3	40	57	11	56	33	0	99	1	14	195	91	94.5	4	<0.00001

NA not aware; The *p*-values of the chi-square test show that the opinions of farmers strongly differ between states

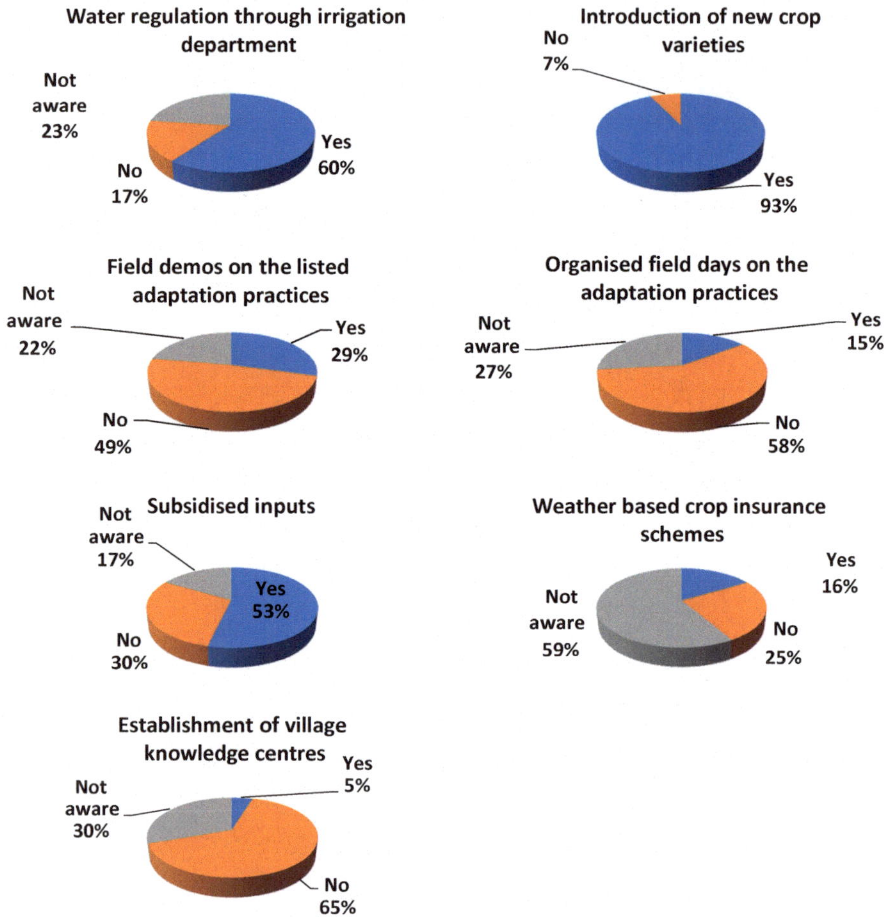

Fig. 5.4 Opinions of farmers of northern region on various government initiatives

falling directly on the crop and interseasonal water accumulation within a year (Iski and Devadoss 2006; Kim and Pang 2009). The summary statistics of all the variables are discussed below:

(a) **Yield (tons per ha)**

The yield data (Table 5.11 and Fig. 5.5) shows an upward trend in all the three states over years. Trend equations fitted to yield data (Table 5.11) show that all the coefficients of trend are strongly significant (1% level). Further, Punjab consistently maintained higher yield during all the 31 years (Fig. 5.5).

(b) **Maximum Temperature** (°C)

The maximum temperature had an overall average of 33.82 °C (Table 5.12). Punjab had the highest mean maximum temperature of 34.9 °C followed by UP (33.85 °C).

Table 5.11 Descriptive statistics of yield and trend equations (tons/ha)

	N	Mean	Min	Max	Std. Dev.	Trend equation	R^2
Punjab	31	3.72	2.74	4.70	0.58	$Y = 2.7760 + 0.0592^{***}\ t$	0.8574
Haryana	31	2.75	2.22	3.65	0.33	$Y = 2.4132 + 0.0208^{***}\ t$	0.3279
UP	31	2.09	1.24	3.33	0.55	$Y = 1.4847 + 0.0380^{***}\ t$	0.3992
Overall	93	2.85	1.24	4.70	0.84	$Y = 1.4637 + 0.0393^{***}\ t + 1.6306D1 + 0.6521D2$	0.8305

***Significant at 1% level; $D1$ and $D2$ are dummy variables representing Punjab and Haryana states

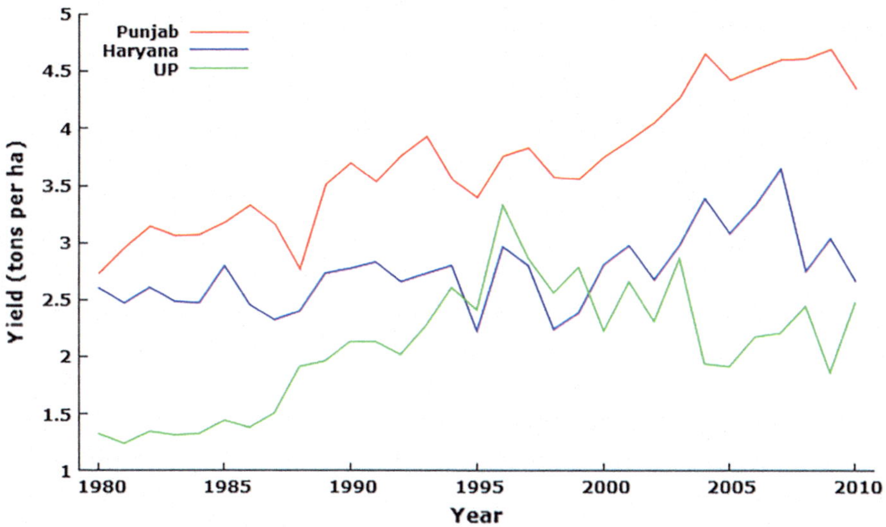

Fig. 5.5 Trend in yield of rice in northern region of India

Table 5.12 Descriptive statistics of maximum temperature (°C)

	N	Mean	Minimum	Maximum	Std. Dev.
Punjab	31	34.90	34.09	36.60	0.58
Haryana	31	32.72	31.65	35.49	0.76
UP	31	33.85	31.33	35.05	0.76
Overall	93	33.82	33.11	35.64	0.51

Also the variability in maximum temperature is least for Punjab. Its maximum temperature is consistently higher than those of Haryana and UP in most of the years (Fig. 5.6).

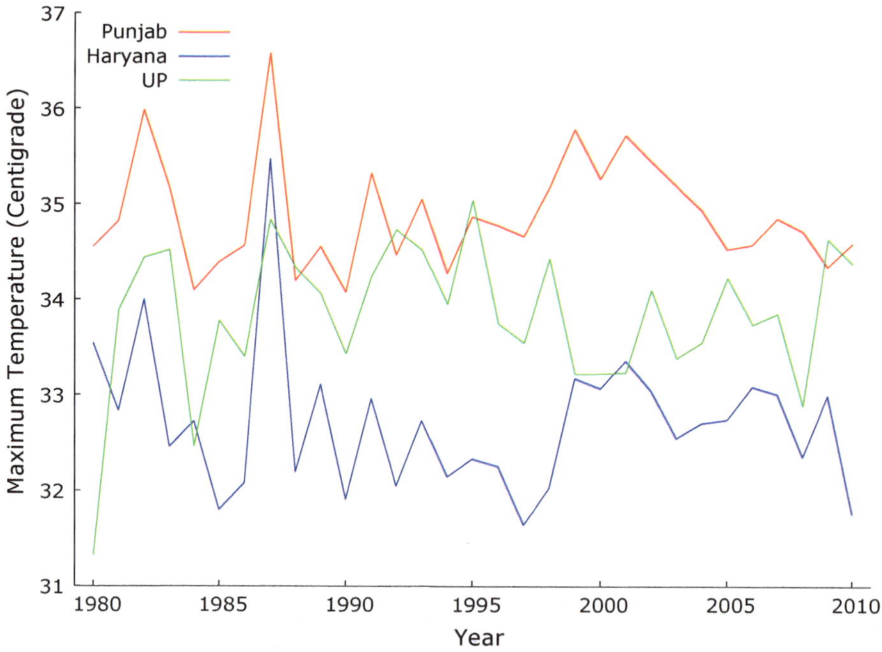

Fig. 5.6 Trend in maximum temperature in northern region of India

Table 5.13 Descriptive statistics of minimum temperature (°C)

	N	Mean	Minimum	Maximum	Std. Dev.
Punjab	31	24.14	23.20	26.36	0.66
Haryana	31	22.94	21.80	23.97	0.49
UP	31	25.10	23.27	26.28	0.63
Overall	93	24.06	23.17	25.21	0.45

(c) **Minimum Temperature** (°C)

The minimum temperature data shows that Uttar Pradesh has the highest average minimum temperature compared to other two states. Its average temperature is 25.1 °C (Table 5.13). It is consistently higher than those of Punjab and Haryana in most of the years. Haryana has the lowest minimum temperature in all the 31 years (Fig. 5.7).

(d) **Average Temperature** (°C)

Table 5.14 shows that the overall average temperature of the region was 28.94 °C. Punjab had the maximum average temperature of 29.52 °C, and Haryana had the lowest average temperature of 27.83 °C. The standard deviation varies between 0.62 and 0.70 °C across the three states. Figure 5.8 shows that the average temperature of Haryana was consistently lower than those of Punjab and Uttar Pradesh.

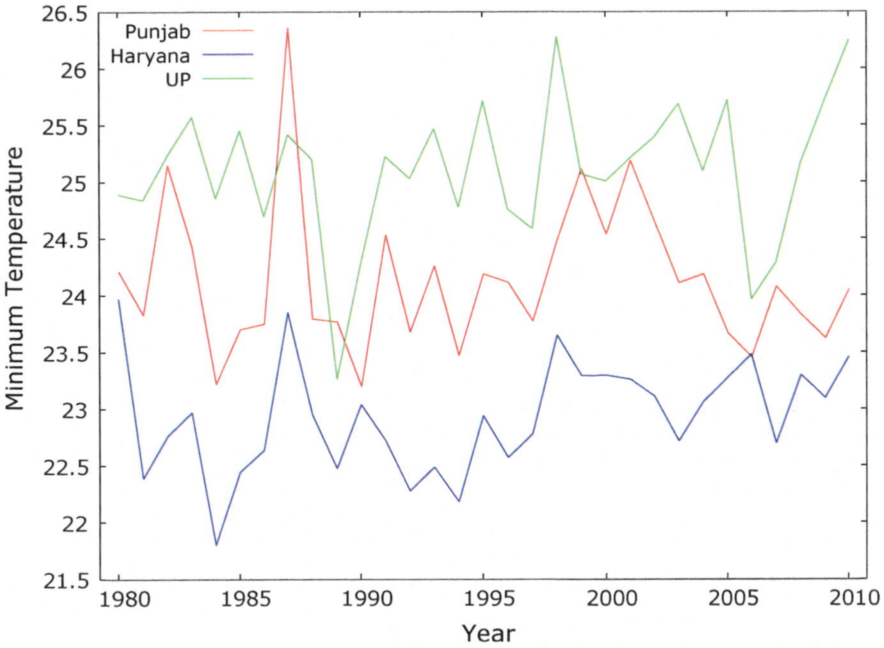

Fig. 5.7 Trend in minimum temperature in northern region of India

Table 5.14 Descriptive statistics of average temperature (°C)

	N	Mean	Minimum	Maximum	Std. Dev.
Punjab	31	29.52	28.65	31.48	0.62
Haryana	31	27.83	26.73	29.73	0.63
UP	31	29.48	27.30	30.67	0.70
Overall	93	28.94	27.13	31.48	0.97

(e) **Rainfall (mm)**

Annual rainfall was highest for UP followed by Haryana. Even though the average rainfall for Punjab is the least (482 mm), it has highest annual variability in rainfall with 216 mm (Table 5.15). Rainfall in UP is consistently higher than those of Punjab and Haryana during most of the years (Fig. 5.9), and similarly the rainfall of Punjab is lower than those of the other two states in most of the years. Thus, we find that while Punjab has higher temperature than those of Haryana and UP, it has lower rainfall compared to these two states.

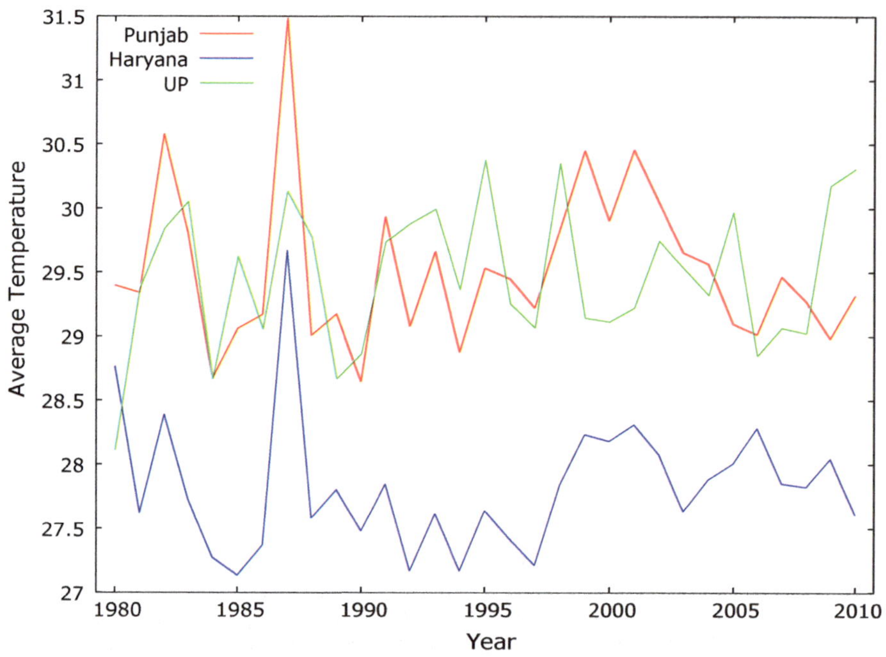

Fig. 5.8 Trend in average temperature in northern region of India

Table 5.15 Descriptive statistics of average rainfall (mm)

	Mean	Minimum	Maximum	Std. Dev.
Punjab	482	217	1270	216
Haryana	600	271	1177	193
UP	991	671	1387	174
Overall	691	489	1085	132

5.5.2 *Test of Normality of Yield Data*

The normality of yield data was tested using Shapiro–Wilk W test and Lilliefors test. The tests showed that the null hypothesis that the data is normal cannot be rejected. The computed value of Shapiro–Wilk W test was 0.9748 with a p-value of 0.0685. Since this value is greater than 0.05, the null hypothesis was not rejected. Similarly, the test statistic for Lilliefors test was 1.2875 with a p-value of 0.08 which is also greater than 0.05. So both tests did not reject the null hypothesis that yield data is normal.

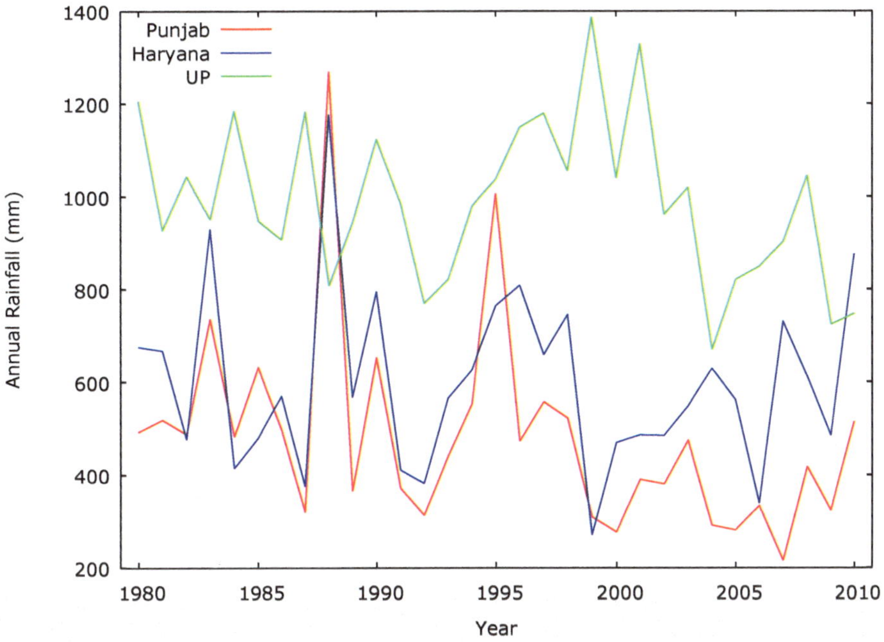

Fig. 5.9 Trend in annual rainfall in northern region of India

Table 5.16 Panel root test results using Fisher's tests

Variable	Test statistic for individual effects and no trend	*p*-value	Decision	Test statistic for individual effect with trend	*p*-value	Decision
Yield	10.5839	0.1021	Accept H_0	14.6465	0.0232**	Reject H_0
Rainfall	27.6310	0.0001***	Reject H_0	27.6310	0.0000***	Reject H_0
Maximum temperature	27.6310	0.0001***	Reject H_0	27.6310	0.0000***	Reject H_0
Minimum temperature	27.6310	0.0001***	Reject H_0	27.6310	0.0000***	Reject H_0

Note *** and **denote significant at 1 and 5% levels

5.5.3 Panel Root Test: Fisher's Tests

Since we have data on rice yield from three states for 31 years, we have a panel data set. So it must be subjected to panel root testing before modelling. As stated in Chap. 4, Fisher's panel root test was applied and the results are presented in Table 5.16.

Table 5.17 Estimation results of Just–Pope production function

Mean function	Linear		Quadratic	
	Coefficient	S. error	Coefficient	S. error
Constant (β_0)	3.2099**	1.5152	−8.1441	35.0335
Trend (β_1)	0.0442***	0.0038	0.0456***	0.0037
Max$T$ (β_2)	0.0706	0.0576	4.2178*	2.3954
Min$T$ (β_3)	−0.1653**	0.0653	−5.3732***	1.7349
Rain$T$ (β_4)	−0.0001	0.0002	0.0099*	0.0057
Max$T$2 (β_5)	–	–	−0.1520***	0.0541
Min$T$2 (β_6)	–	–	−0.1114*	0.0577
Rain$T$2 (β_7)	–	–	0.0000	0.0000
Max$T$Min$T$ (β_8)	–	–	0.2876***	0.0933
Max$T$Rain$T$ (β_9)	–	–	−0.0010***	0.0002
Min$T$Rain$T$ (β_{10})	–	–	0.0011***	0.0003
Punjab (β_{11})	1.3733***	0.1396	1.5113***	0.1582
Haryana (β_{12})	0.4389**	0.1720	0.5099***	0.1774
Variance function				
Constant	0.5333	6.2219	5.0761	5.4721
Time	0.0448**	0.0226	0.0089**	0.0210
Max$T$	−0.3251	0.2813	−0.5245*	0.2692
Min$T$	0.2715	0.3107	0.4007	0.3068
Rain$T$	0.0011	0.0008	0.0004	0.0008
Model statistics				
LL	−19.1135		−9.7478	
AIC	62.2270		55.4955	
BIC	92.6181		101.0823	

Note ***, ** and *denote significant at 1, 5 and 10% levels

The test results confirmed that all the variables in panel data do not have unit roots and hence they are stationary.

5.5.4 Estimation of Just–Pope Production Function

As discussed in the methodology chapter, to account for both mean and variance in yield, Just–Pope production function (Just and Pope 1978 and 79) with yield as dependent variable and maximum and minimum temperatures, rainfall, their squares and interaction terms and dummy variables for states were used as independent variables. Table 5.17 gives the results of the estimated empirical models.

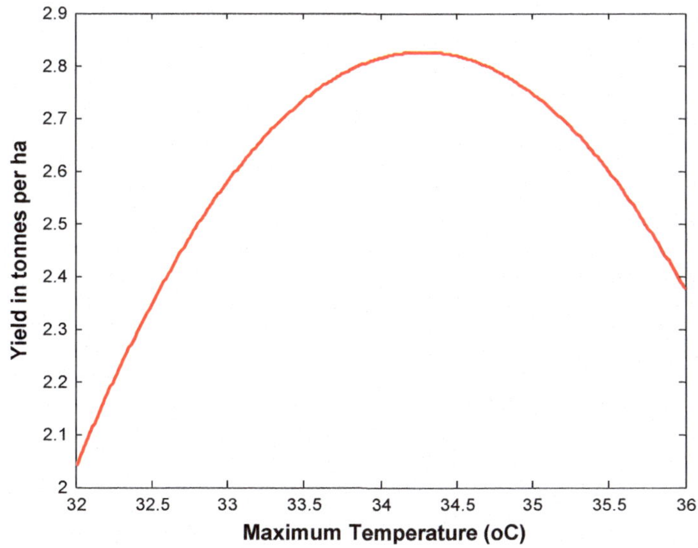

Fig. 5.10 Nonlinear relationship between yield and maximum temperature

The quadratic model has the least value for AIC and so is selected. Table 5.17 shows that maximum temperature has a significant positive effect on rice yield, whereas minimum temperature has very strong significant negative effect. Similarly, rainfall has significant positive effect on yield. All the square terms (except that of rainfall) and all interaction terms are significant. The significance of the two square terms and all the three interaction terms imply that nonlinear effects of climate variables are significant and so important (Figs. 5.10, 5.11 and 5.12). The coefficients of the two dummies for Punjab and Haryana states are positive and strongly significant implying that the mean yields between the states differ statistically. The coefficient for Punjab is 1.5113 and that of Haryana is 0.5099. This means that, on the average, difference in mean yield (per ha) between these two states is 1.0014 assuming all other variables fixed. The positive and significant coefficient for trend implies the significance of technological progress in rice production such as improved irrigation methods, use of high yielding varieties and better control of pests and diseases over the period under study.

With respect to variance function, coefficients of trend and maximum temperature are significant. The coefficient of trend is positive implying that it is a risk-increasing variable. This means that variability in yield increases over years. But we have already seen that the coefficient of trend in the mean yield function is positive and significant which implies that mean yield increases over time. So, we conclude that over years increase in mean yield increases variability in yield also. These results coincide with the findings of Palanisami et al. (2011), Ranganathan (2009), Isik and Devadoss (2006) and Anderson and Hazell (1987). Since the coefficient of maximum temperature is negative, it is a risk-decreasing variable for rice yield. So, increase

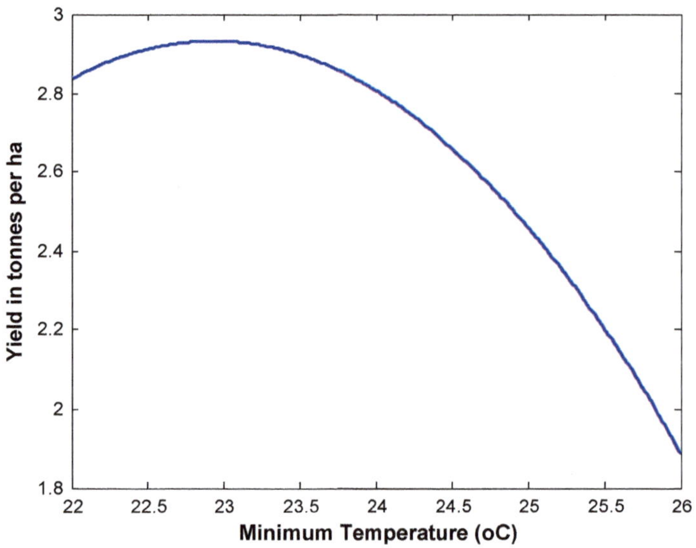

Fig. 5.11 Nonlinear relationship between yield and minimum temperature

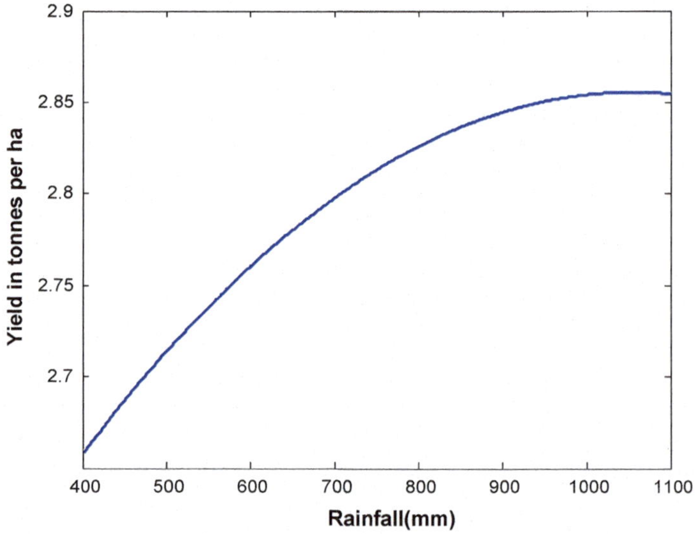

Fig. 5.12 Nonlinear relationship between yield and rainfall

Table 5.18 Elasticities of climatic variables at means

Yield function	Climate variable	Elasticity
Mean	Maximum temperature	1.6240
	Minimum temperature	−2.1055
	Rainfall	0.0803
Variance	Maximum temperature	−17.7406
	Minimum temperature	9.6402
	Rainfall	0.2532

in maximum temperature decreases the yield variability. Coefficients of minimum temperature and rainfall though positive are not statistically significant.

The elasticities for all climatic variables at mean levels are presented in Table 5.18 for both mean and variance functions. The elasticity for the maximum temperature is 1.624 implying that 1% increase in maximum temperature will induce an increase of 1.62% in mean yield. Similarly, the elasticity coefficient for minimum temperature is −2.1055 and so 1% increase in minimum temperature will reduce the mean yield by 2.11%. The elasticity coefficient for rainfall is 0.0803 which is very small, and so 1% increase in rainfall will induce an increase of 0.08% in mean yield. When all the three variables are simultaneously increased by 1%, we can expect a small decrease of 0.4% in yield.

For the variance function, the elasticity coefficient of maximum temperature is −17.74 implying a reduction in variability in yield of 17.7 for 1% increase in maximum temperature. Similar interpretations can be made for the elasticities of minimum temperature and rainfall, even though their coefficients are not significant. All together, we can expect a decrease of 7.85% in yield when all the three variables are increased by 1%. So we conclude that all the three variables put together are risk decreasing in this sense.

The mean function can be used to compute the optimal levels of climate variables which will give maximum yield. Using differential calculus, it can be shown that the optimal levels of Max$T$, Min$T$ and Rain$T$ are obtained by solving the following set of simultaneous equations:

$$\frac{\partial y}{\partial(\mathrm{Max}T)} = 2\beta_5\mathrm{Max}T + \beta_8\mathrm{Min}T + \beta_9\mathrm{Rain}T = -\beta_2$$
$$\frac{\partial y}{\partial(\mathrm{Min}T)} = \beta_8\mathrm{Max}T + 2^*\beta_6\mathrm{Min}T + \beta_{10}\mathrm{Rain}T = -\beta_3$$
$$\frac{\partial y}{\partial(\mathrm{Rain}T)} = \beta_9\mathrm{Max}T + \beta_{10}\mathrm{Rain}T + 2\beta_7\mathrm{Rain}T = -\beta_4$$

Solving these equations, the optimal levels are obtained as Max$T$ = 36.86 °C; Min$T$ = 26.61 °C and Rain$T$ = 639.3 mm.

Table 5.19 Climate change impacts on percentage of mean yield and yield variability

Scenario	Period	Percentage change in				
		*T*max	*T*min	Rainfall	Mean yield	Yield variability
RCP4.5	2021–2050	1.3232	4.9525	17.5011	−6.87	28.70
	2071–2100	6.5661	11.5596	11.7217	−12.73	−2.08
RCP8.5	2021–2050	0.7417	5.5802	25.1846	−8.52	47.01
	2071–2100	9.7056	19.3691	44.8659	−21.42	25.90

5.5.5 Climate Change Scenarios and Rice Yield Changes

The estimated elasticities given in Table 5.18 were used to estimate the impacts of climate change on mean yield and yield variability of yield. For this purpose, two climate change scenarios, viz. RCP4.5 and RCP8.5 (as discussed in Chap. 3), were considered. The results are presented in Table 5.19.

The table shows that, as per RCP4.5 and RCP8.5 scenarios, there will be substantial reduction in percentages of rice yield during the periods 2021–50 and 2071–2100. Under both scenarios, percentage reduction in the later period will be more. The percentage reduction in mean yield varies from 6.87 to 8.52 during the first period as per the two scenarios, and it varies from 12.73 to 21.42% during the second period.

Yield variability as estimated by RCP8.5 scenario is more than that predicted by RCP4.5 scenario for both periods. The first scenario predicts an increase of 28.7% yield variability during 2021–2050, while the second scenario predicts an increase of 47% for the same period. For the second period, RCP4.5 predicts a small reduction in yield variability of 2%, whereas RCP8.5 scenario predicts an increase in yield variability of 25.9%.

5.6 Summary

Punjab and Uttar Pradesh farmers perceive untimely rain followed by incidence of pests and diseases and depletion of groundwater as most important climate change events, while, for farmers of Haryana, depletion in groundwater is the foremost concern. Farmers ranked decline in crop yield as the most important climate change shock. All farmers in the three states have restricted rice-growing season to one. In addition, all the farmers in Punjab and Haryana have changed crop variety and planting dates as mitigation strategies.

Farmers are well aware of introduction of new crop varieties, good understanding of water regulation by government and on the subsidised inputs. But, not many farmers participated in the field demos, weather-based cropping insurance schemes and other government initiatives. This conclusion calls for formulating suitable popularisation mechanisms by the government.

Based on the historical data, yield of rice shows an upward trend in all the three states over years. Punjab had the highest mean maximum temperature followed by UP, while UP had the highest minimum temperature. Annual rainfall in UP was higher than those of the other two states. Maximum temperature has a significant positive effect on rice yield, whereas minimum temperature has very strong significant negative effect. Similarly, rainfall has significant positive effect on yield. Also, there is strong evidence of nonlinear effects of climate variables on yield. The two climate change scenarios RCP4.5 and RCP8.5 predict substantial reduction in rice yield during the periods 2021–50 and 2071–2100. Yield variability as estimated by the second scenario (RCP8.5) is more than that predicted by the first scenario (RCP4.5) for both periods.

Appendix

Table 5.20 Factors influencing farmers perception of climate change—Garrett ranking: Punjab

Factors	1	2	3	4	5	6	7	8	9	10	11	12	13	14	15	16	Total
Drought	1	1	1	2	3	6	4	8	9	6	2	5	6	13	14	19	100
Hailstorm	0	0	0	2	0	1	2	3	1	5	4	12	13	21	22	14	100
Flood	0	1	0	0	0	2	0	0	5	0	2	5	13	16	23	33	100
AD	0	5	3	6	6	4	7	3	6	8	10	7	18	4	6	7	100
IPD	13	19	12	19	5	8	5	2	4	1	2	1	4	2	1	2	100
IFM	3	6	7	3	13	11	13	4	3	6	8	10	5	5	1	2	100
UR	13	19	20	7	6	6	5	3	1	3	3	4	3	4	2	1	100
IW	8	15	14	14	8	5	4	2	4	4	2	4	3	6	4	3	100
TFH	5	15	12	16	15	13	5	4	5	4	0	0	1	1	3	1	100
TFL	1	8	19	13	19	9	4	7	3	4	2	3	4	0	3	1	100
SSDI	0	1	4	5	7	5	11	11	14	10	10	9	4	6	2	1	100
CCP	0	0	1	2	6	8	11	14	9	10	11	16	3	6	0	3	100
CLA	0	0	0	1	4	6	13	12	15	9	12	6	5	10	5	2	100
CFI	0	1	0	1	2	10	6	12	11	18	9	10	8	3	7	2	100
CF	1	0	0	0	2	2	8	8	9	11	23	7	10	3	7	9	100
DGW	55	9	7	9	4	4	2	7	1	1	0	1	0	0	0	0	100
Total	100	100	100	100	100	100	100	100	100	100	100	100	100	100	100	100	1600
Garrett score	86	76	70	65	61	58	55	52	48	45	42	39	35	30	24	14	

Table 5.21 Rankings of factors representing climate change using Garrett ranking: Punjab

Climate change event	Rank																Total score	Average score	Rank
	1	2	3	4	5	6	7	8	9	10	11	12	13	14	15	16			
	Total score																		
Drought	86	76	70	130	183	348	220	416	432	270	84	195	210	390	336	266	3712	37.12	14
Hailstorm	0	0	0	130	0	58	110	156	48	225	168	468	455	630	528	196	3172	31.72	15
Flood	0	76	0	0	0	116	0	0	240	0	84	195	455	480	552	462	2660	26.6	16
AD	0	380	210	390	366	232	385	156	288	360	420	273	630	120	144	98	4452	44.52	12
IPD	1118	1444	840	1235	305	464	275	104	192	45	84	39	140	60	24	28	6397	63.97	2
IFM	258	456	490	195	793	638	715	208	144	270	336	390	175	150	24	28	5270	52.7	7
UR	1118	1444	1400	455	366	348	275	156	48	135	126	156	105	120	48	14	6314	63.14	3
IW	688	1140	980	910	488	290	220	104	192	180	84	156	105	180	96	42	5855	58.55	6
TFH	430	1140	840	1040	915	754	275	208	240	180	0	0	35	30	72	14	6173	61.73	4
TFL	86	608	1330	845	1159	522	220	364	144	180	84	117	140	0	72	14	5885	58.85	5
SSDI	0	76	280	325	427	290	605	572	672	450	420	351	140	180	48	14	4850	48.5	8
CCP	0	0	70	130	366	464	605	728	432	450	462	624	105	180	0	42	4658	46.58	9
CLA	0	0	0	65	244	348	715	624	720	405	504	234	175	300	120	28	4482	44.82	10
CFI	0	76	0	65	122	580	330	624	528	810	378	390	280	90	168	28	4469	44.69	11
CF	86	0	0	0	122	116	440	416	432	495	966	273	350	90	168	126	4080	40.8	13
DGW	4730	684	490	585	244	232	110	364	48	45	0	39	0	0	0	0	7571	75.71	1

Table 5.22 Factors influencing farmers perception of climate change—Garrett ranking: Uttar Pradesh

Climate change event	Rank																Total
	1	2	3	4	5	6	7	8	9	10	11	12	13	14	15	16	
Drought	1	0	1	4	8	6	5	2	4	3	5	6	7	12	35	1	100
Hailstorm	48	17	11	4	3	3	3	3	2	2	4	0	0	0	0	0	100
Flood	9	35	14	9	1	2	8	5	4	4	5	0	4	0	0	0	100
AD	0	0	2	0	1	2	8	17	17	10	10	7	7	11	7	1	100
IPD	12	13	5	15	10	16	17	9	3	0	0	0	0	0	0	0	100
IFM	0	0	0	0	0	0	0	0	0	0	0	1	0	5	15	79	100
UR	18	22	27	3	8	7	8	3	0	2	0	2	0	0	0	0	100
IW	9	8	12	16	5	11	10	6	5	3	2	0	1	4	6	2	100
TFH	0	0	0	1	0	2	2	10	6	11	24	21	18	4	1	0	100
TFL	0	0	0	0	0	0	2	3	4	9	9	27	26	16	4	0	100
SSDI	0	0	0	2	1	7	17	22	27	11	8	1	3	0	1	0	100
CCP	0	0	0	0	0	0	1	0	2	8	15	14	13	13	24	10	100
CLA	0	0	0	0	0	3	1	7	17	18	9	10	12	20	2	1	100
CFI	0	0	2	23	18	17	8	6	5	6	2	6	6	1	0	0	100
CF	3	5	24	19	31	15	1	0	0	0	1	0	0	1	0	0	100
DGW	0	0	2	4	14	9	9	7	4	13	6	5	3	13	5	6	100
Total	100	100	100	100	100	100	100	100	100	100	100	100	100	100	100	100	1600
Garrett score	86	76	70	65	61	58	55	52	48	45	42	39	35	30	24	14	

Table 5.23 Rankings of factors representing climate change using Garrett ranking: Uttar Pradesh

Climate change event	Rank																Total score	Average score	Rank
	1	2	3	4	5	6	7	8	9	10	11	12	13	14	15	16			
	Total score																		
Drought	86	0	70	260	488	348	275	104	192	135	210	234	245	360	840	14	3861	38.61	13
Hailstorm	4128	1292	770	260	183	174	165	156	96	90	168	0	0	0	0	0	7482	74.82	1
Flood	774	2660	980	585	61	116	440	260	192	180	210	0	140	0	0	0	6598	65.98	3
AD	0	0	140	0	61	116	440	884	816	450	420	273	245	330	168	14	4357	43.57	10
IPD	1032	988	350	975	610	928	935	468	144	0	0	0	0	0	0	0	6430	64.3	5
IFM	0	0	0	0	0	0	0	0	0	0	0	39	0	150	360	1106	1655	16.55	16
UR	1548	1672	1890	195	488	406	440	156	0	90	0	78	0	0	0	0	6963	69.63	2
IW	774	608	840	1040	305	638	550	312	240	135	84	0	35	120	144	28	5853	58.53	6
TFH	0	0	0	65	0	116	110	520	288	495	1008	819	630	120	24	0	4195	41.95	11
TFL	0	0	0	0	0	0	110	156	192	405	378	1053	910	480	96	0	3780	37.8	14
SSDI	0	0	0	130	61	406	935	1144	1296	495	336	39	105	0	24	0	4971	49.71	8
CCP	0	0	0	0	0	0	55	0	96	360	630	546	455	390	576	140	3248	32.48	15
CLA	0	0	0	0	0	174	55	364	816	810	378	390	420	600	48	14	4069	40.69	12
CFI	0	0	140	1495	1098	986	440	312	240	270	84	234	210	30	0	0	5539	55.39	7
CF	258	380	1680	1235	1891	870	55	0	0	0	42	0	0	30	0	0	6441	64.41	4
DGW	0	0	140	260	854	522	495	364	192	585	252	195	105	390	120	84	4558	45.58	9

Chapter 6
Climate Change Impacts on Rice Yield in Eastern Region of India

K. Palanisami, Krishna Reddy Kakumanu, Udaya Sekhar Nagothu and C. R. Ranganathan

6.1 Introduction

The East Region for the present study covers four states, viz. Bihar, Odisha, Chhattisgarh and West Bengal, and the districts selected from these States were Jehanabad, Bhadrak, Raipur and Bankura, respectively. As stated in the methodology chapter, the present study is based on (i) primary data: farm survey among 100 farmers from Bihar, 500 farmers from Odisha, 70 farmers from Chhattisgarh and 151 farmers from West Bengal and (ii) secondary data: time series data (1982–2010) on productivity (yield) of rice and climate variables from the three districts belonging to Bihar, Odisha and West Bengal (since Chhattisgarh is a newly formed State, it was excluded). The location of the selected districts in eastern region is shown in Fig. 6.1.

6.2 Socio-economic Profiles of the Farmers

From the total of 821 farmers surveyed, 790 were males and 31 were females. Out of them, all the 100 farmers in Bihar and 51 farmers in West Bengal have irrigated crop-based farming. Also, 200 farmers in Odisha and 100 farmers in West Bengal follow crop with supplemental irrigation typology. Thus, in West Bengal, both irrigation-based and crop with supplemental irrigation cropping typologies are prevalent. All the farmers in Chhattisgarh and 100 farmers in Odisha have dry-land-based typologies.

The average age of the farmers in the four States, viz. Bihar, Odisha, Chhattisgarh and West Bengal is 45, 49, 44 and 43 years, respectively. About 12% of the farmers in the region have no formal education, 32% have studied up to 5th standard, 46% up to primary level and the rest, 11% have studied up to secondary level (Fig. 6.2) and 56% of the farmers from Odisha have education up to primary level. Thus, the education level of the farmers in the region is low.

K. Palanisami et al., *Climate Change and Future Rice Production in India*, India Studies in Business and Economics,
https://doi.org/10.1007/978-981-13-8363-2_6

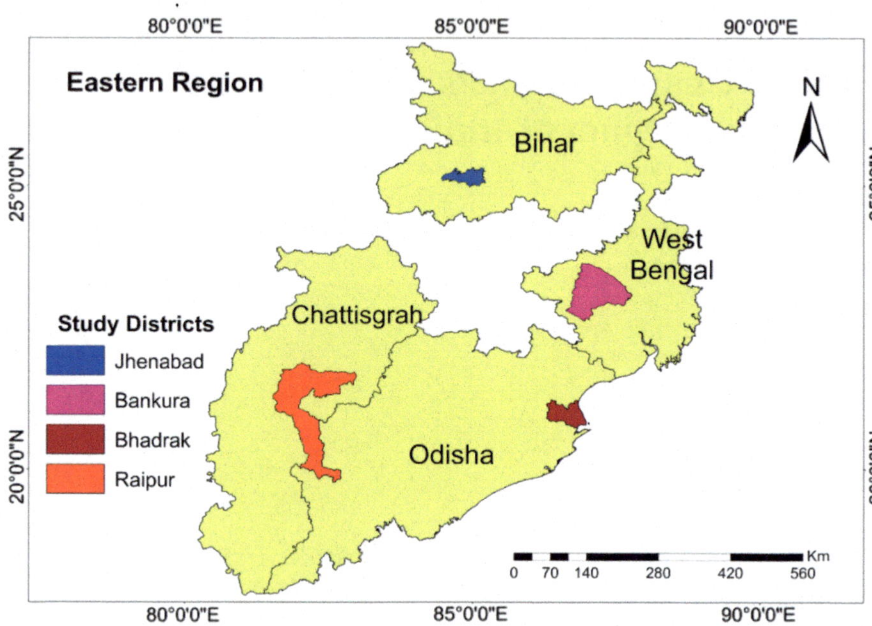

Fig. 6.1 Study regions

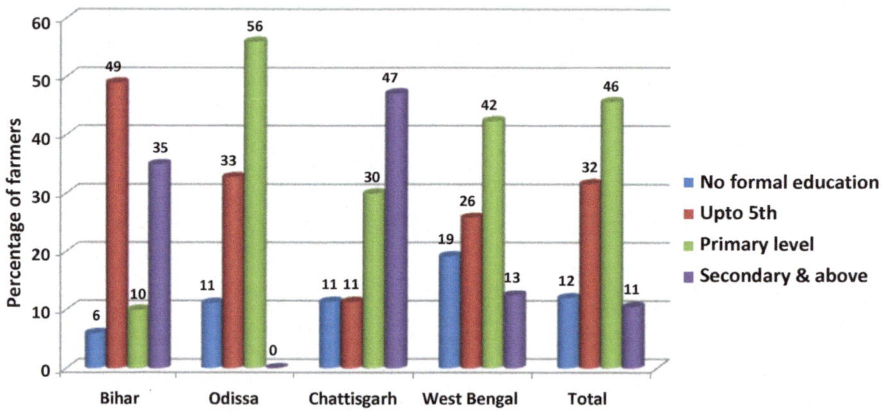

Fig. 6.2 Educational status of sample farmers

6.3 Farm Information

Farmers in the study area possess land and wells. They use different sources of irrigation like tanks, canals and wells, and they invest in digging borewells. In Bihar and Odisha, they raise rice crop only during the kharif season. Farmers in Chhattisgarh raise rice crop in kharif and rabi seasons, and in West Bengal, farmers raise rice in all three seasons, viz. kharif, rabi and summer. The farmers in West Bengal also raise other crops. Overall, farmers from the four states derive income mainly from rice during kharif season. Though farmers in West Bengal could get income from rice crop in kharif, they could get income from other crops also during all the seasons.

(a) **Physical assets**:

(i) **Land owned**: Farmers in Bihar own wet, irrigated and dry lands, whereas farmers in Odisha own only wet and irrigated land. They do not have dry lands. Chhattisgarh farmers possess only irrigated lands. Farmers belonging to West Bengal own irrigated and dry lands. The summary statistics is given in Table 6.1.

The average size of land owned by farmers in the four states are, respectively, 0.77, 1.06, 2.34 and 0.85 ha. The mean area under irrigation varies from 0.94 to 2.34 ha across the four states with the highest area corresponding to Chhattisgarh. But, Chhattisgarh has the highest standard deviation of 2.31 ha of irrigated land. The average area under wetland in Bihar is 0.69 ha while 1.0 ha in Odisha. The average area under dry land is 0.63 ha in Bihar and 0.45 ha in West Bengal.

(ii) **Wells owned**: No farmer possesses open or borewells in Odisha and West Bengal, and farmers in Bihar possess no open wells whereas only four farmers in Chhattisgarh possess open wells. Table 6.2 gives the frequency distribution of number of borewells owned by the farmers in Bihar and Chhattisgarh states.

Table 6.1 Summary statistics of land owned (hectares)

State	Land type	Mean	S.D.	Min	Max	Count
Bihar	Wet	0.69	0.41	0.40	2.02	68
	Irrigated	0.94	0.51	0.40	3.64	100
	Dry	0.63	0.31	0.40	1.62	84
	Overall	0.77	0.45	0.40	3.64	252
Odisha	Wet	1.00	0.57	0.40	3.24	193
	Irrigated	1.11	0.99	0.20	8.09	300
	Overall	1.06	0.85	0.20	8.09	493
Chhattisgarh	Irrigated	2.34	2.31	0.40	14.16	70
West Bengal	Irrigated	1.00	1.25	0.03	12.02	151
	Dry	0.45	0.55	0.04	3.64	59
	Overall	0.85	1.13	0.03	12.02	210

Table 6.2 Distribution of number of borewells owned across states

No. of wells	Number of farmers owning			
	Bihar	Chhattisgarh		Total
	Borewells	Open wells	Borewells	
0	18	66	43	127
1	61	4	27	92
2	12	–	–	12
3	6	–	–	6
4	2	–	–	2
>4	1	–	–	1

Table 6.3 Sources of irrigation in West Bengal

Source of irrigation	No. of farmers
Canal	1
Tank	89
well	3
Well (Village)	7
Well (Village) & canal	47
Well (Own)	1
Well (Own) & canal	1
Well (Village and own) & canal	2
Total	151

It is evident that borewells are common in these two states. The sources of irrigation in West Bengal are shown in Table 6.3.

(iii) **Sources of irrigation**. Farmers have three sources of irrigation, viz. tanks, canals and wells. All farmers in Bihar depend on wells only (Fig. 6.3).

(iv) **Investment in irrigation**: No farmer in all the four states used water-saving technologies like drip and sprinkler irrigation.

(v) **Area (ha) under rice in the region**.

Rice is grown in all the four states during kharif season. The average area under rice per farmer during this season ranged between 0.79 and 1.07 ha. In Chhattisgarh, rice is grown during rabi season also (Tables 6.4 and 6.5).

(b) **Net Income from rice and other crops**

The average net income per ha has a wide variation among the states. During kharif season, it varies between Rs. 79,479 for Chhattisgarh and Rs. 5822 for Bihar. Farmers in all the states except Chhattisgarh derive income from other crops also and West Bengal farmer earns an average income of Rs. 46,493.

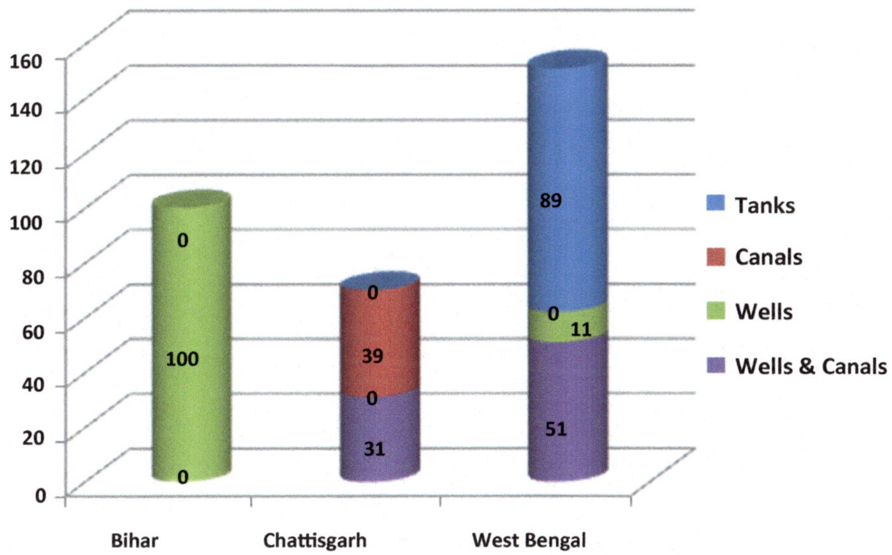

Fig. 6.3 Frequency distribution of sources of irrigation

Table 6.4 Average area (ha) per farmer under rice across the states in eastern region

State	Season	Rice	Other crops
Bihar	Kharif	0.79 (100)	–
Odisha	Kharif	1.06 (500)	–
	Rabi	–	0.23 (40)
Chhattisgarh	Kharif	2.34 (70)	–
	Rabi	0.63 (11)	–
West Bengal	Kharif	1.07 (151)	0.13 (51)
	Rabi	–	0.42 (107)
	Summer	1.38 (51)	0.15 (59)

Figures in brackets denote the number of farmers who raised the particular crop in the particular season

6.4 Climate Change Events, Shocks and Their Effects as Perceived by Farmers

Farmers' perception of the climate change, its effects on agriculture and various adaptation strategies/techniques followed by the farmers are assessed based on the detailed survey. The results are discussed below.

Table 6.5 Net income from rice and other crops

State	Net income (Rs/ha)				
	Rice			Wheat	Other crops
	Kharif	Rabi	Summer	Rabi	
Bihar	5822 (100)	–	–	8194 (100)	2862 (100)
Odisha	17,725 (500)	–	–	–	31,870 (39)
Chhattisgarh	79,479 (70)	82,881(11)	–	–	–
West Bengal	28,533 (151)	–	13,911 (51)	–	46,493 (111)
Overall	23,528 (821)	82,881 (11)	13,911 (51)	8194 (100)	26,759 (250)

Figures in brackets denote the number of farmers who raised the particular crop in the particular season

6.4.1 Perception of Farmers on Climate Change Events

Farmers were asked to rank the 16 events as listed in the methodology chapter according to their perceptions of climate change. To find what the farmers perceive as the most significant effects of climate change, Garrett's ranking technique as described in Chap. 4 was used. The results are given in Tables 6.6 and 6.7. Table 6.7 shows that crop failure is the most important climate change event (Rank 1) perceived by Rice farmers in eastern region. It earned an average Garrett Score of 69.54. As given in Table 6.6, it was given Rank 1 by 500 farmers. The second and third important climate change events are, respectively, depletion of groundwater and drought. Similar analysis was done separately for each state, viz. Bihar, Odisha, Chhattisgarh and West Bengal. The results are presented in Appendix 6.1.

Friedman Rank Test was applied to test for differences between rankings of the climate change events (listed in Table 6.6) by the farmers. The calculated value of Friedman's F statistics was 6214.3 while the table value (corresponding to 15 degrees of freedom at 5% level of significance) is 24.996. This shows that the farmers did not consistently rank the events and the farmers' perceptions of climate change effects differ across the sample chosen. The same conclusion was arrived at by applying Kendall's coefficient of concordance. Similar analysis was done for individual states, viz. Bihar, Odisha, Chhattisgarh and West Bengal. The study results confirm that there is no consistency in the rankings of the events by the farmers. Table 6.8 gives the rankings of the climate change events state-wise.

Table 6.8 shows that the rankings of the climate change events are not uniform across the states. So, to check for agreement in rankings, Kendall's Coefficient of Concordance was computed and tested for its significance. The computed value of Kendall's Coefficient, W was 0.4493 (with number of objects $n = 16$ and number of judges $m = 3$) with a χ^2 value of 26.9559 (p-value $= 0.0291$). Since the p-value is less than 0.05, we reject the null hypothesis and conclude that the states are not concordant in ranking the climate change events. That is, the states differ in ranking the climate change events. The statewise information on the factors influencing and the ranking of the factors are given in Appendix (Tables 6.21, 6.22, 6.23, 6.24, 6.25, 6.26, 6.27, 6.28).

Table 6.6 Factors influencing farmers perception of climate change—Garrett ranking: eastern region

Climate change event	Rank																Total
	1	2	3	4	5	6	7	8	9	10	11	12	13	14	15	16	
Drought	173	7	200	4	303	3	14	16	16	8	8	16	27	26	0	0	821
Hailstorm	0	0	0	36	55	36	24	0	0	1	92	23	29	18	6	501	821
Flood	0	1	9	16	14	25	20	11	4	0	200	70	0	0	451	0	821
AD	41	19	9	24	21	21	35	19	15	114	9	119	71	204	100	0	821
IPD	33	19	7	88	7	511	5	13	12	6	22	17	25	21	25	10	821
IFM	5	5	5	3	15	18	10	11	6	9	22	39	146	407	114	6	821
UR	29	171	66	117	13	11	111	0	301	0	0	2	0	0	0	0	821
IW	7	31	125	148	24	19	307	121	21	12	4	0	2	0	0	0	821
TFH	27	37	44	33	97	21	117	409	16	16	4	0	0	0	0	0	821
TFL	0	3	5	11	9	83	7	14	17	220	101	0	200	0	0	151	821
SSDI	2	8	306	16	225	24	101	32	34	28	28	13	4	0	0	0	821
CCP	0	1	0	303	1	1	3	6	205	16	27	35	46	62	92	23	821
CLA	1	1	11	9	13	18	18	32	24	147	230	226	16	5	0	70	821
CFI	0	0	0	2	1	4	7	14	18	178	22	238	237	43	22	35	821
CF	500	3	5	0	6	10	20	95	29	24	37	16	13	27	11	25	821
DGW	3	515	29	12	16	17	21	28	103	41	16	6	6	8	0	0	821
Total	821	821	821	822	820	822	820	821	821	820	822	820	822	821	821	821	13,136
Garrett score	86	76	70	65	61	58	55	52	48	45	42	39	35	30	24	14	

Table 6.7 Rankings of factors representing climate change using Garrett ranking: eastern region

Climate change event	Rank																Total	Average	Rank
	1	2	3	4	5	6	7	8	9	10	11	12	13	14	15	16			
	Total score																		
Drought	14,878	532	14,000	260	18,483	174	770	832	768	360	336	624	945	780	0	0	53,742	65.4592	3
Hailstorm	0	0	0	2340	3355	2088	1320	0	0	45	3864	897	1015	540	144	7014	22,622	27.5542	16
Flood	0	76	630	1040	854	1450	1100	572	192	0	8400	2730	0	0	10,824	0	27,868	33.94397	14
AD	3526	1444	630	1560	1281	1218	1925	988	720	5130	378	4641	2485	6120	2400	0	34,446	41.95615	10
IPD	2838	1444	490	5720	427	29,638	275	676	576	270	924	663	875	630	600	140	46,186	56.25579	8
IFM	430	380	350	195	915	1044	550	572	288	405	924	1521	5110	12,210	2736	84	27,714	33.75639	15
UR	2494	12,996	4620	7605	793	638	6105	0	14,448	0	0	78	0	0	0	0	49,777	60.62972	5
IW	602	2356	8750	9620	1464	1102	16,885	6292	1008	540	168	0	70	0	0	0	48,857	59.50914	6
TFH	2322	2812	3080	2145	5917	1218	6435	21,268	768	720	168	0	0	0	0	0	46,853	57.06821	7
TFL	0	228	350	715	549	4814	385	728	816	9900	4242	0	7000	0	0	2114	31,841	38.78319	12
SSDI	172	608	21,420	1040	13,725	1392	5555	1664	1632	1260	1176	507	140	0	0	0	50,291	61.25579	4
CCP	0	76	0	19,695	61	58	165	312	9840	720	1134	1365	1610	1860	2208	322	39,426	48.02192	9
CLA	86	76	770	585	793	1044	990	1664	1152	6615	9660	8814	560	150	0	980	33,939	41.33861	11
CFI	0	0	0	130	61	232	385	728	864	8010	924	9282	8295	1290	528	490	31,219	38.02558	13
CF	43,000	228	350	0	366	580	1100	4940	1392	1080	1554	624	455	810	264	350	57,093	69.5408	1
DGW	258	39,140	2030	780	976	986	1155	1456	4944	1845	672	234	210	240	0	0	54,926	66.90134	2

Table 6.8 Ranking of climate change events state-wise

Climate change event	Bihar	Odisha	CGH	WB	Eastern region
Drought	1	4	1	12	3
Hailstorm	11	16	11	6	16
Flood	4	14	12	15	14
AD	5	13	13	5	10
IPD	14	6	4	4	8
IFM	12	15	14	8	15
UR	2	8	2	2	5
IW	7	7	3	3	6
TFH	6	9	5	1	7
TFL	9	11	6	16	12
SSDI	8	3	7	7	4
CCP	15	5	15	14	9
CLA	10	10	16	11	11
CFI	16	12	10	13	13
CF	13	1	8	9	1
DGW	3	2	9	10	2

6.4.2 *Effects of Shocks*

Farmers were questioned on their perception about the effects of climatic shocks. They were asked to rank the seven effects of climatic shocks, as described in methodology chapter. Farmers ranked declining crop yield as the most important shock as out of 821 farmers, 647 of them ranked this factor as number one (Tables 6.9 and 6.10). The second and third ranks are given to the loss of income and food security/shortage, respectively. This is understandable because the decline in crop yield will imply loss of income and food shortage.

6.4.3 *Mitigation Strategies Used by Farmers*

Farmers apply several mitigation strategies to tide over the negative impacts of climate change. Table 6.11 provides a summary. It shows that 83% of the farmers in the region have restricted rice growing season to one. About 88% of the farmers (54% in total) in Odisha have changed crop variety and planting dates. These two strategies were not followed by farmers in other states. Only 22% of the farmers have left their land fallow. Also, 82% of the farmers in the region borrowed money from relatives/others and only 38% received assistance from the government. All other strategies are followed by less than 20% of the farmers.

Table 6.9 Rankings of effects of climatic shocks by farmers

Factors	Rank							Total no. of farmers
	1	2	3	4	5	6	7	
DCY	647	88	61	14	0	11	0	821
LA	0	23	37	223	228	135	175	821
LI	50	475	131	89	51	23	2	821
FIS	4	118	229	61	284	108	17	821
DL	24	15	32	161	165	149	275	821
DC	51	42	245	230	72	66	115	821
DH	43	60	88	41	22	331	236	821
Total	819	821	823	819	822	823	820	5747
Garrett score	79	66	57	50	43	34	21	

Table 6.10 Garrett ranking of effects of climatic shocks

Factors	Rank									Rank
	1	2	3	4	5	6	7	Total score	Average score	
	Total score									
DCY	51,113	5808	3477	700	0	374	0	61,472	74.87454	1
LA	0	1518	2109	11,150	9804	4590	3675	32,846	40.00731	5
LI	3950	31,350	7467	4450	2193	782	42	50,234	61.18636	2
FIS	316	7788	13,053	3050	12,212	3672	357	40,448	49.26675	3
DL	1896	990	1824	8050	7095	5066	5775	30,696	37.38855	7
DC	4029	2772	13,965	11,500	3096	2244	2415	40,021	48.74665	4
DH	3397	3960	5016	2050	946	11,254	4956	31,579	38.46407	6

Table 6.11 Frequency distribution of mitigation strategies followed by farmers

Strategies	Bihar	Odisha	Chhattisgarh	WB	Total	Percentage
Did nothing	100	0	0	0	100	12
Changed crop variety	0	441	0	0	441	54
Changed planting date	0	441	0	0	441	54
Left land fallow	100	14	0	64	178	22
Sold part of land	0	15	0	7	22	3
Leased out part of land	0	153	0	14	167	20
Restricted number of seasons for rice growing to 1	100	488	0	95	683	83
Sold livestock	0	1	0	87	88	11

(continued)

Table 6.11 (continued)

Strategies	Bihar	Odisha	Chhattisgarh	WB	Total	Percentage
Shifted to other crops	0	0	0	47	47	6
Borrowed money from relatives/others	100	493	0	82	675	82
Received assistance from government/NGOs	0	139	70	99	308	38
Less food consumption	0	0	0	66	66	8
Shifted to non-farm employment	19	1	11	90	121	15
Outmigration to cities	0	44	0	6	50	6

6.4.4 Opinions of Farmers on Government Initiatives

From time to time, both Central and State governments implement several mitigations and long-term measures to offset the negative effects of climate change on agriculture. With a view to know the awareness of these measures among the farmers, they were questioned on six government initiatives (as described in the methodology chapter). They were asked to state their replies as either 'yes', or 'no' or 'not aware'. Figure 6.4 gives a pictorial representation of the awareness of these initiatives. About 95% of the farmers are well aware of subsidized inputs (like fertilizers, drip and machinery), 73% know about the introduction of new crop varieties, 70% have knowledge on water-based crop insurance schemes and 62% of the farmers know about water regulation by the government through irrigation department. Only 16% of the farmers know about the establishment of village knowledge centres and 14% only are aware of field demos on the listed adaptation practices. The last two government initiatives are very important for the farmers to know about latest technologies and so suitable mechanisms should be formed by the government to popularize these two initiatives. Table 6.12 provides a summary of the opinions on government initiatives, state-wise.

Chi-square test was applied to check if the opinions differ across the states and the results show that the opinions of the farmers on government initiatives differ strongly among the three states. This may be due to the government initiatives depending on local natural resources availability and other factors.

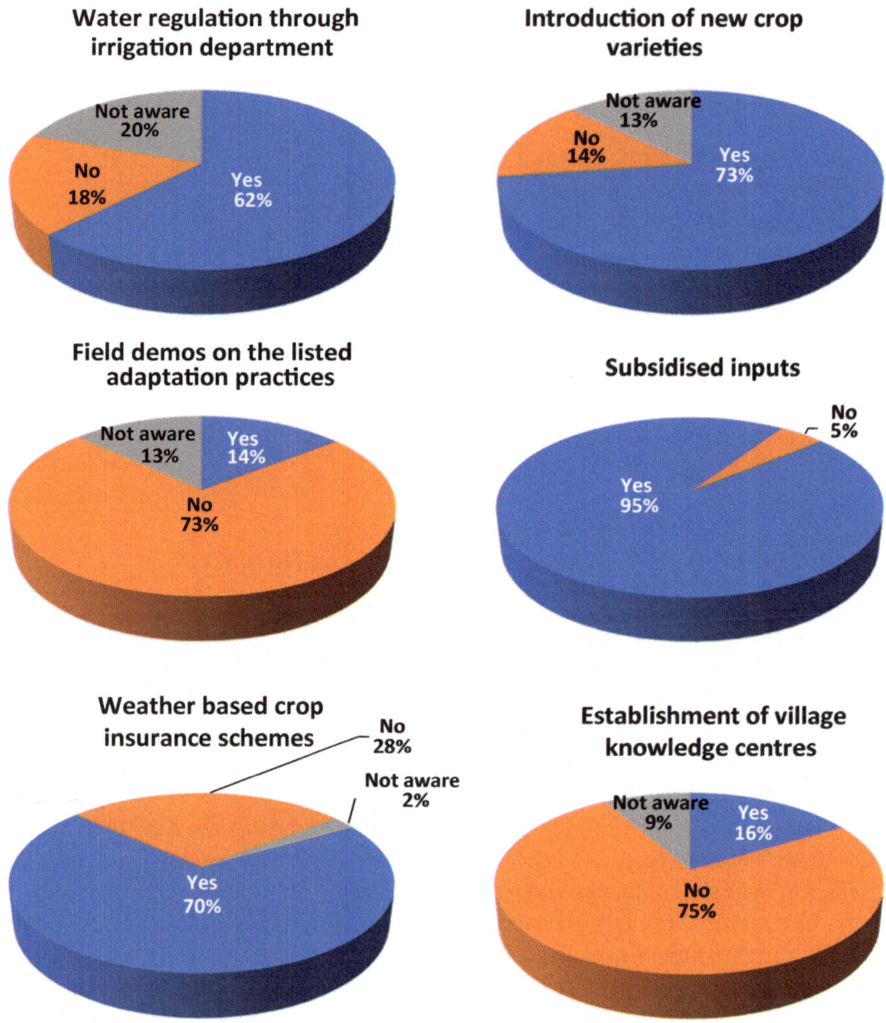

Fig. 6.4 Opinions of farmers of eastern region on various government initiatives

6.5 Analysis of the Effect of Climate Change on Rice Yield

6.5.1 Data

The objective of the study is to estimate the effects of climate variables on the yield of rice crop using historical data. Hence, we used a panel data set of yield and climate variables pertaining to 29 years from 1982 to 2010. The data on temperature corresponds to the growing season of rice variety in the respective states and the time series data on rainfall was collected for the entire year to account for rainfall falling directly on the crop and interseasonal water accumulation within a year (Iski and

Table 6.12 Farmers' opinions on government initiatives

Govt. initiatives	Bihar			Odisha			Chhattisgarh			West Bengal			Total			Chi-square	df	*p*-value
	Yes	No	NA	Yes	No	NA	Yes	No	NA	Yes	No	NA	Yes	No	NA			
Water regulation through irrigation department	0	0	100	388	112	0	70	0	0	54	36	61	512	148	161	800.16	6	<0.00001
Introduction of new crop varieties	0	0	100	498	2	0	70	0	0	29	118	4	597	120	104	1451.6	6	<0.00001
Field demos on the listed adaptation practices	0	100	0	1	399	100	70	0	0	47	96	8	118	595	108	584.63	6	<0.00001
Organized field days on the adaptation practices	0	100	0	0	400	100	70	0	0	42	103	6	112	603	106	602.01	6	<0.00001
Subsidized inputs (fertilizer, drip and machinery)	100	0	0	499	1	0	70	0	0	114	37	0	783	38	0	165.58	6	<0.00001
Weather-based crop insurance schemes	0	100	0	500	0	0	70	0	0	4	128	19	574	228	19	848.52	6	<0.00001
Establishment of village knowledge centres, climate cell, etc.	0	100	0	133	367	0	0	70	0	1	81	69	134	618	69	412.38	6	<0.00001

NA not aware; The *p*-values of the chi-square test show that the opinions of farmers strongly differ between states

Table 6.13 Descriptive statistics of yield and trend equations (tons/ha)

	N	Mean	Min	Max	Std. Dev.	Trend equation	R^2
Bihar	29	1.2144	0.681	1.600	0.244	$Y = 1.0183 + 0.0131^{**}$ T	0.21
Odisha	29	1.3094	0.737	1.795	0.278	$Y = 1.0513 + 0.0172^{***}$ T	0.28
West Bengal	29	2.1569	1.018	3.346	0.475	$Y = 1.4333 + 0.0482^{***}$ T	0.75
Overall	87	1.560	0.681	3.346	0.547	–	–

** and *** denotes significane at 5% and 1% level

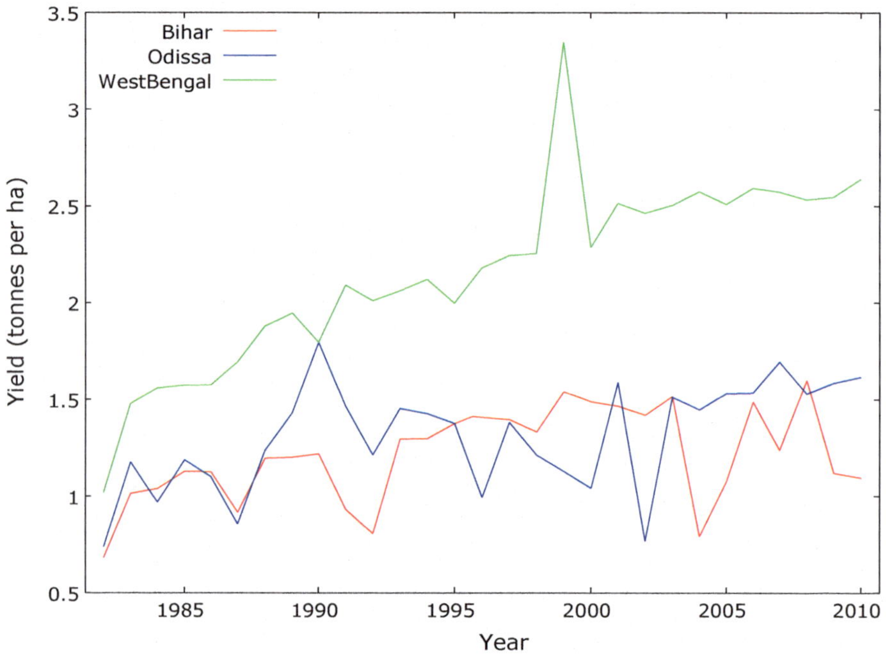

Fig. 6.5 Trend in yield of rice in eastern region of India

Devadoss 2006; Kim and Pang 2009). Since Chhattisgarh State was formed during 2000 only, this State was excluded from panel data analysis and only three states, viz. Bihar, Odisha and West Bengal were included. The summary statistics of all the variables are discussed below.

(a) **Yield (tons per ha)**

The yield data (Table 6.13; Fig. 6.5) shows an upward trend in all the three states over years, and yield for West Bengal was greater than that of the other two states in all the 29 years.. Trend equations fitted to yield data (Table 6.13) show that the coefficient of trend for all the three states is statistically significant.

(b) **Maximum Temperature** (°C)

The maximum temperature had an overall average of 32.19 °C. Odisha had the highest mean maximum temperature of 32.5 °C followed by Bihar (32.3 °C). Also, the variability in maximum temperature is least for Bihar and highest for West Bengal (Table 6.14; Fig. 6.6).

(c) **Minimum Temperature** (°C)

The minimum temperature data shows that Odisha has the highest average minimum temperature compared to the other two states. Its average minimum temperature is 22.02 °C. It is consistently higher than those of the other two states in most of

Table 6.14 Descriptive statistics of maximum temperature (°C)

	N	Mean	Minimum	Maximum	Std. Dev.
Bihar	29	32.325	31.814	32.903	0.298
Odisha	29	32.483	31.750	34.250	0.634
West Bengal	29	32.106	29.750	33.670	0.858
Overall	87	32.190	29.750	34.250	1.121

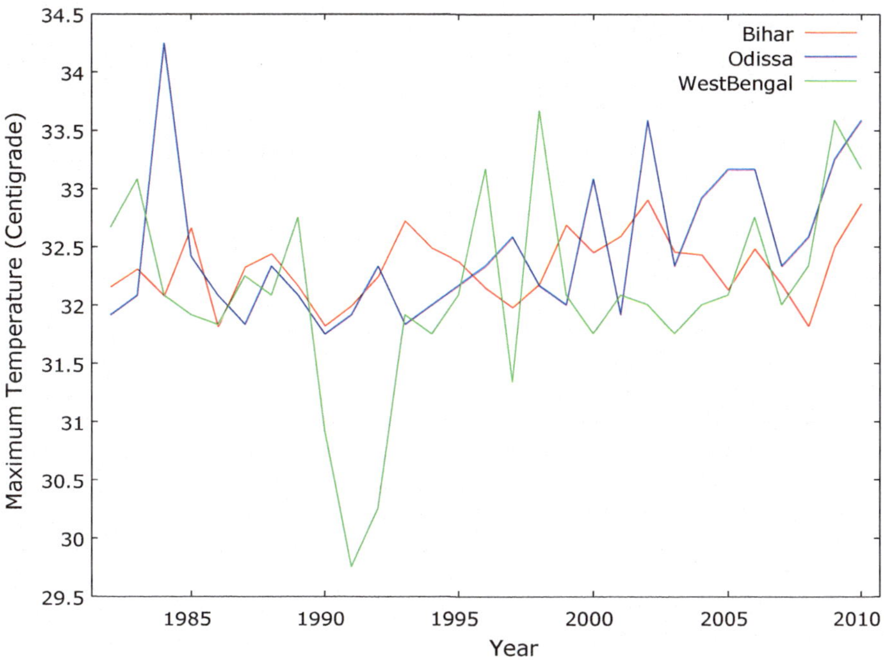

Fig. 6.6 Trend in maximum temperature in eastern region of India

Table 6.15 Descriptive statistics of minimum temperature (°C)

	N	Mean	Minimum	Maximum	Std. Dev.
Bihar	29	19.936	19.305	20.481	0.363
Odisha	29	22.026	19.250	23.750	1.022
West Bengal	29	21.138	19.333	22.167	0.768
Overall	87	21.033	19.250	23.750	1.148

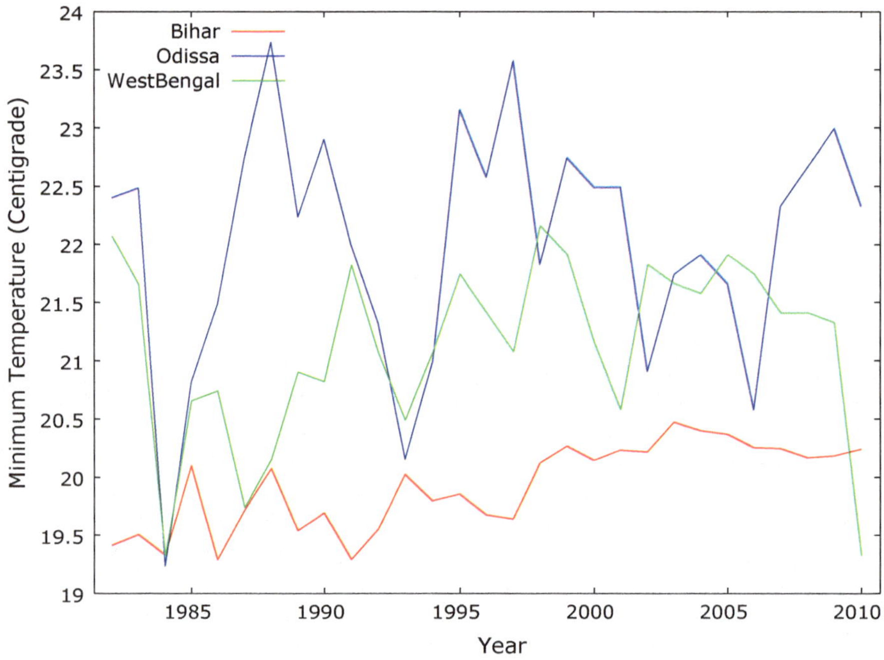

Fig. 6.7 Trend in minimum temperature in eastern region of India

the years. Bihar had the lowest minimum temperature with an average of 19.93 °C (Table 6.15; Fig. 6.7).

(d) **Rainfall** (mm)

Average annual rainfall was the highest for Odisha (1488 mm) followed by West Bengal (1410 mm) (Table 6.16). Also, variability in rainfall was the highest for Odisha with a standard deviation of 310 mm. Both Bihar and West Bengal had the same variability in rainfall, but the average rainfall of Bihar was least (885.79 mm). Rainfall for Bihar was consistently lower than that of the other two states in most of the years (Fig. 6.8).

Table 6.16 Descriptive statistics of rainfall (mm)

	N	Mean	Minimum	Maximum	Std. Dev.
Bihar	29	885.79	518.10	1472.5	226.87
Odisha	29	1488.0	922.60	2052.6	309.54
West Bengal	29	1401.3	917.0	1954.0	227.91
Overall	87	1258.4	518.10	2052.6	369.25

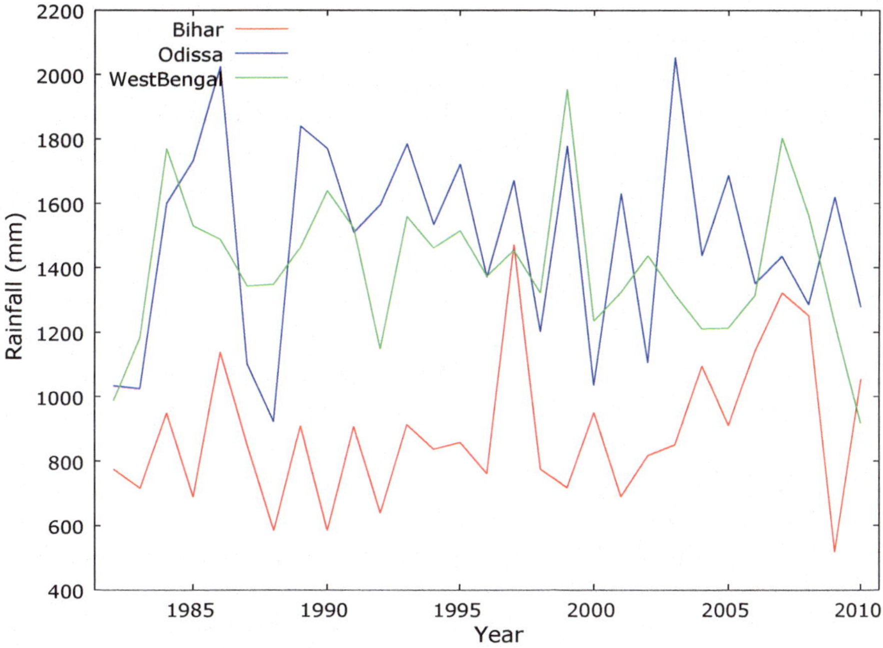

Fig. 6.8 Trend in annual rainfall in eastern region of India

6.5.2 Test of Normality of Yield Data

The normality of yield data was tested and the tests showed that the data slightly deviates from normality and hence, the outliers were removed to have a normally distributed data.

Table 6.17 Panel root test results using Fisher's tests

Variable	Test statistic for individual effects and no trend	*p*-value	Decision	Test statistic for individual effect with trend	*p*-value	Decision
Yield	20.6380	0.0021	Reject H_0	26.2447	0.0002	Reject H_0
Maximum temperature	18.6992	0.0047	Reject H_0	19.2068	0.0038	Reject H_0
Minimum temperature	10.9396	0.0093	Reject H_0	17.3331	0.0081	Reject H_0
Rainfall	27.631	0.0001	Reject H_0	24.2215	0.0005	Reject H_0

6.5.3 Panel Root Test: Fisher's Tests

The data on rice yield and climate variables for three States for 29 years is available and was subjected to panel root testing before modelling. As stated in the methodology Chap. 3, Fisher's panel root test was applied.

The test results confirmed (Table 6.17) that all the variables in panel data do not have unit roots and hence, they are stationary.

6.5.4 Estimation of Just–Pope Production Function

As discussed in the methodology chapter, to account for both mean and variance in yield, Just–Pope production function (Just and Pope 1978 & 79) with yield as the dependent variable and maximum and minimum temperatures, rainfall, their squares and interaction terms and dummy variables for states were used as independent variables. Table 6.18 gives the results of the estimated empirical models.

The quadratic model has the least value for AIC and hence selected. Table 6.18 shows that maximum temperature has a significant positive effect on rice yield, whereas rainfall has a significant negative effect. The square term corresponding to maximum temperature and its interaction term with rainfall are significant. The significance of these terms implies that nonlinear effects of maximum temperature are significant and so important (Fig. 6.9). The coefficient of the two dummies for Bihar and Odisha are negative and strongly significant. The positive and significant coefficient for trend implies the significance of technological progress in rice production such as improved irrigation methods, use of high-yielding varieties and better control of pests and diseases over the period under study.

In the case of variance function, coefficient of maximum temperature alone is significant and also positive. So, it is risk-increasing variable. This means that vari-

Table 6.18 Estimation results of Just–Pope production function

Mean function	Linear		Quadratic	
	Coefficient	S. error	Coefficient	S. error
Constant (β_0)	0.4945	2.3423	−73.7162	91.0495
Trend (β_1)	0.0248***	0.0044	0.0282***	0.0047
Max$T$ (β_2)	−0.0040	0.0572	3.2601**	1.3023
Min$T$ (β_3)	0.0410	0.0550	2.8695	3.2449
Rain$T$ (β_4)	0.0002	0.0001	−0.0119*	0.0072
Max$T$2 (β_5)	–	–	−0.0164***	0.0043
Min$T$2 (β_6)	–	–	0.0227	0.0272
Rain$T$2 (β_7)	–	–	0.0000	0.0000
Max$T$Min$T$ (β_8)	–	–	−0.1230	0.0987
Max$T$Rain$T$ (β_9)	–	–	0.0003**	0.0001
Min$T$Rain$T$ (β_{10})	–	–	0.0001	0.0001
Bihar (β_{11})	−0.7763***	0.1305	−0.8840***	0.1419
Odisha (β_{12})	−0.8466***	0.1196	−0.8832***	0.1008
Variance function				
Constant	−45.9696***	8.4349	−55.3047	10.4527
Time	−0.0045	0.0280	−0.0032	0.0303
Max$T$	1.2445***	0.1968	1.5594***	0.2878
Min$T$	0.1063	0.2502	0.0637	0.2463
Rain$T$	0.0007	0.0006	0.0007	0.0007
Model statistics				
LL	1.175168		7.7028	
AIC	20.650		20.5943	

** and *** denotes significane at 5% and 1% level

ability in yield increases with an increase in maximum temperature. Other variables are not statistically significant.

The elasticities for all climatic variables at mean levels are presented in Table 6.19 for both mean and variance functions. The elasticity of the three climate variables is all less than 1, and so they are inelastic. The elasticity of the maximum temperature for the mean function is −0.4629 implying that 1% increase in maximum temperature will induce a decrease of 0.46% in mean yield. Similarly, interpretations can be given to the other two elasticities. When all the three variables are simultaneously increased by 1%, we can expect a very small decrease of 0.04% in yield.

For the variance function, the elasticity coefficient of maximum temperature is −50.1963 implying an increase in variability in yield of 5.02 for 1% increase in maximum temperature. Similar meanings can be given for the elasticities of minimum temperature and rainfall. All together, we can expect an increase of 7.2% in yield

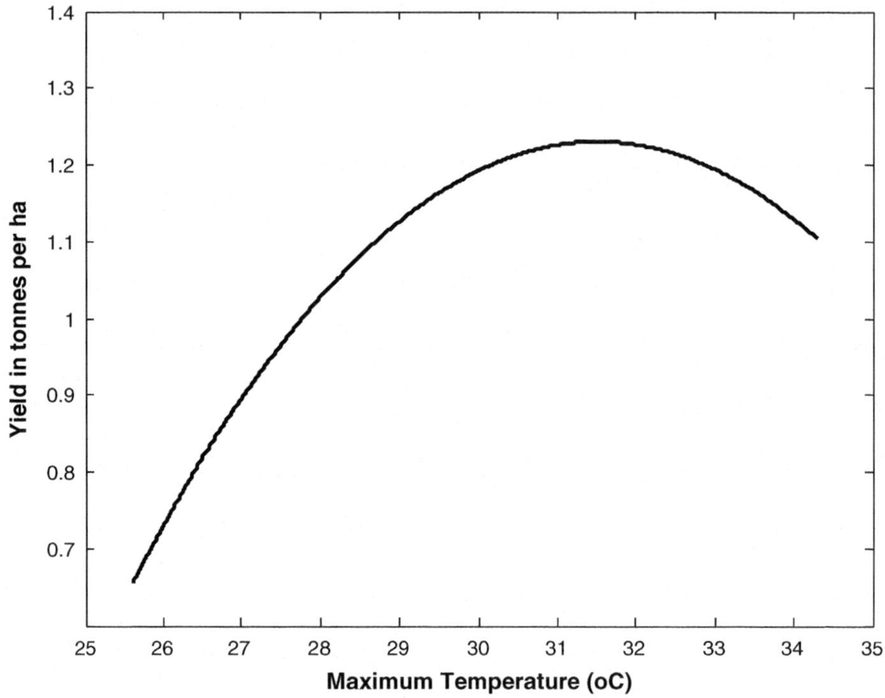

Fig. 6.9 Nonlinear relationship between yield and maximum temperatures

Table 6.19 Elasticities of climatic variables at means

Yield function	Climate variable	Elasticity
Mean	Maximum temperature	−0.4629
	Minimum temperature	0.1644
	Rainfall	0.1827
Variance	Maximum temperature	5.0196
	Minimum temperature	1.3394
	Rainfall	0.8393

variability when all the three variables are increased by 1% simultaneously. So, we conclude that all the three variables, put together are risk increasing in this sense.

The mean function can be used to compute the optimal levels of climate variables which will give maximum yield. Using differential calculus, it can be shown that the optimal levels of Max*T*, Min*T* and Rain*T* are obtained by solving the following set of simultaneous equations:

$$\frac{\partial y}{\partial(\mathrm{Max}T)} = 2\beta_5\mathrm{Max}T + \beta_8\mathrm{Min}T + \beta_9\mathrm{Rain}T = -\beta_2$$

$$\frac{\partial y}{\partial(\text{Min}T)} = \beta_8\text{Max}T + 2 * \beta_6\text{Min}T + \beta_{10}\text{Rain}T = -\beta_3$$

$$\frac{\partial y}{\partial(\text{Rain}T)} = \beta_9\text{Max}T + \beta_{10}\text{Rain}T + 2\beta_7\text{Rain}T = -\beta_4$$

Solving these equations, the optimal levels are obtained as Max$T$ = 31.9 °C; Min$T$ = 20.4 °C and Rain$T$ = 1043 mm.

6.5.5 Climate Change Scenarios and Rice Yield Changes

The estimated elasticities given in Table 6.19 were used to estimate the impacts of climate change on mean yield and yield variability of yield. For this purpose, two climate change scenarios, viz. RCP4.5 and RCP8.5 (as discussed in Chap. 3) were considered (Table 6.20).

Table 6.20 shows that, as per RCP4.5 and RCP8.5 scenarios, there will be a marginal increase in percentages of rice yield during the periods 2021–50 and 2071–2100. Under both scenarios, percentage of increase in the later period will be more. The percentage increase in mean yield varies from 1.40 to 1.63 during the first period as per the two scenarios, and it varies from 0.96 to 2.36% during the second period as per the two scenarios.

Yield variability as estimated by RCP8.5 scenario is more than that predicted by RCP4.5 scenario for both periods. The first scenario predicts an increase of 35% yield variability during 2021–2050 while the second scenario predicts an increase of 41% for the same period. For the second period, RCP4.5 predicts a higher increase in yield variability of 83%, whereas RCP8.5 scenario predicts an increase in yield variability of 139%.

Thus, we conclude that climate change may not have adverse effect on the mean rice yield in eastern region but yield variability may be substantial.

Table 6.20 Climate change impacts on percentage of mean yield and yield variability

Scenario	Period	Percentage change in			Percentage change in	
		T_{max}	T_{min}	Rainfall	Mean rice yield	Yield variability
RCP4.5	2021–2050	3.36	7.10	9.82	1.40	34.63
	2071–2100	9.79	14.42	17.09	0.96	82.79
RCP8.5	2021–2050	3.97	8.86	11.00	1.63	41.04
	2071–2100	15.80	26.33	29.28	2.36	139.14

6.6 Summary

Eastern region farmers perceive crop failure followed by the depletion of groundwater and drought as the most important climate change events. As in the case of northern region, farmers of this region also ranked decline in crop yield as the most important climate change shock. As far as mitigation strategies, more than 80% of the farmers in the region have restricted rice growing season to one and they borrowed money from their relatives/friends.

About 95% of the farmers are well aware of subsidized inputs as they have good knowledge about the introduction of new crop varieties and weather-based crop insurance schemes. Only the small number of farmers knows about the establishment of village knowledge centres and aware of field demos on the listed adaptation practices. These two government initiatives are very important for the farmers to update their knowledge about the latest technologies and so suitable steps should be taken by the government to popularize these two initiatives.

Based on historical data, yield of rice generally shows an upward trend in all the states over the years, although there was a decline in yield in some years. Odisha had the highest maximum temperature, minimum temperature and annual rainfall compared to the other two states. Maximum temperature has a significant positive effect on rice yield, whereas rainfall has significant negative effect. The square term corresponding to maximum temperature and its interaction term with rainfall is significant implying nonlinear effects of climatic variables on yield. The two climate change scenarios RCP4.5 and RCP8.5 predict a marginal increase in percentages of rice yield during the periods 2021–50 and 2071–2100. Also, yield variability as estimated by the second scenario is more than that predicted by the first scenario for both periods.

Appendix

.

Table 6.21 Factors influencing farmers' perception of climate change–Garrett ranking: Bihar

Factors	Rank																Total
	1	2	3	4	5	6	7	8	9	10	11	12	13	14	15	16	
Drought	99	1	0	0	0	0	0	0	0	0	0	0	0	0	0	0	100
Hailstorm	0	0	0	0	0	0	0	0	0	1	22	23	29	18	6	1	100
Flood	0	1	9	16	14	25	20	11	4	0	0	0	0	0	0	0	100
AD	0	4	8	17	13	13	20	14	11	0	0	0	0	0	0	0	100
IPD	0	0	0	0	0	0	0	0	0	0	13	13	18	21	25	10	100
IFM	0	0	0	0	0	0	0	0	0	3	11	23	26	17	14	6	100
UR	0	70	20	6	4	0	0	0	0	0	0	0	0	0	0	0	100
IW	1	1	10	12	5	10	7	18	19	12	4	0	1	0	0	0	100
TFH	0	0	8	10	17	10	13	9	15	16	2	0	0	0	0	0	100
TFL	0	3	5	11	9	13	7	14	17	20	1	0	0	0	0	0	100
SSDI	0	3	6	8	15	5	16	13	13	18	3	0	0	0	0	0	100
CCP	0	1	0	0	0	0	0	0	0	3	14	9	7	21	22	23	100
CLA	0	1	9	8	10	11	7	13	14	21	4	2	0	0	0	0	100
CFI	0	0	0	0	0	0	0	0	0	0	8	16	8	11	22	35	100
CF	0	0	0	0	0	1	0	0	3	6	17	14	11	12	11	25	100
DGW	0	15	25	12	13	12	10	8	4	0	1	0	0	0	0	0	100
Total	100	100	100	100	100	100	100	100	100	100	100	100	100	100	100	100	1600
Garrett score	86	76	70	65	61	58	55	52	48	45	42	39	35	30	24	14	

Table 6.22 Rankings of factors representing climate change using Garrett ranking: Bihar

Factors	Rank																Total	Average	Rank
	1	2	3	4	5	6	7	8	9	10	11	12	13	14	15	16			
	Total score																		
Drought	8514	76	0	0	0	0	0	0	0	0	0	0	0	0	0	0	8590	85.9	1
Hailstorm	0	0	0	0	0	0	0	0	0	45	924	897	1015	540	144	14	3579	35.79	11
Flood	0	76	630	1040	854	1450	1100	572	192	0	0	0	0	0	0	0	5914	59.14	4
AD	0	304	560	1105	793	754	1100	728	528	0	0	0	0	0	0	0	5872	58.72	5
IPD	0	0	0	0	0	0	0	0	0	0	546	507	630	630	600	140	3053	30.53	14
IFM	0	0	0	0	0	0	0	0	0	135	462	897	910	510	336	84	3334	33.34	12
UR	0	5320	1400	390	244	0	0	0	0	0	0	0	0	0	0	0	7354	73.54	2
IW	86	76	700	780	305	580	385	936	912	540	168	0	35	0	0	0	5503	55.03	7
TFH	0	0	560	650	1037	580	715	468	720	720	84	0	0	0	0	0	5534	55.34	6
TFL	0	228	350	715	549	754	385	728	816	900	42	0	0	0	0	0	5467	54.67	9
SSDI	0	228	420	520	915	290	880	676	624	810	126	0	0	0	0	0	5489	54.89	8
CCP	0	76	0	0	0	0	0	0	0	135	588	351	245	630	528	322	2875	28.75	15
CLA	0	76	630	520	610	638	385	676	672	945	168	78	0	0	0	0	5398	53.98	10
CFI	0	0	0	0	0	0	0	0	0	0	336	624	280	330	528	490	2588	25.88	16
CF	0	0	0	0	0	58	0	0	144	270	714	546	385	360	264	350	3091	30.91	13
DGW	0	1140	1750	780	793	696	550	416	192	0	42	0	0	0	0	0	6359	63.59	3

Table 6.23 Factors influencing farmers perception of climate change—Garrett ranking: Odisha

Factors	Rank																Total
	1	2	3	4	5	6	7	8	9	10	11	12	13	14	15	16	
Drought	0	0	200	0	300	0	0	0	0	0	0	0	0	0	0	0	500
Hailstorm	0	0	0	0	0	0	0	0	0	0	0	0	0	0	0	500	500
Flood	0	0	0	0	0	0	0	0	0	0	200	0	0	0	300	0	500
AD	0	0	0	0	0	0	0	0	0	100	0	100	0	200	100	0	500
IPD	0	0	0	0	0	500	0	0	0	0	0	0	0	0	0	0	500
IFM	0	0	0	0	0	0	0	0	0	0	0	0	100	300	100	0	500
UR	0	0	0	100	0	0	100	0	300	0	0	0	0	0	0	0	500
IW	0	0	0	100	0	0	300	100	0	0	0	0	0	0	0	0	500
TFH	0	0	0	0	0	0	100	400	0	0	0	0	0	0	0	0	500
TFL	0	0	0	0	0	0	0	0	0	200	100	0	200	0	0	0	500
SSDI	0	0	300	0	200	0	0	0	0	0	0	0	0	0	0	0	500
CCP	0	0	0	300	0	0	0	0	200	0	0	0	0	0	0	0	500
CLA	0	0	0	0	0	0	0	0	0	100	200	200	0	0	0	0	500
CFI	0	0	0	0	0	0	0	0	0	100	0	200	200	0	0	0	500
CF	500	0	0	0	0	0	0	0	0	0	0	0	0	0	0	0	500
DGW	0	500	0	0	0	0	0	0	0	0	0	0	0	0	0	0	500
Total	500	500	500	500	500	500	500	500	500	500	500	500	500	500	500	500	8000
Garrett score	86	76	70	65	61	58	55	52	48	45	42	39	35	30	24	14	

Table 6.24 Rankings of factors representing climate change using Garrett ranking: Odisha

Factors	Rank																Total	Average	Rank
	1	2	3	4	5	6	7	8	9	10	11	12	13	14	15	16			
	Total score																		
Drought	0	0	14,000	0	18,300	0	0	0	0	0	0	0	0	0	0	0	32,300	64.6	4
Hailstorm	0	0	0	0	0	0	0	0	0	0	0	0	0	0	0	7000	7000	14	16
Flood	0	0	0	0	0	0	0	0	0	0	8400	0	0	0	7200	0	15,600	31.2	14
AD	0	0	0	0	0	0	0	0	0	4500	0	3900	0	6000	2400	0	16,800	33.6	13
IPD	0	0	0	0	0	29,000	0	0	0	0	0	0	0	0	0	0	29,000	58	6
IFM	0	0	0	0	0	0	0	0	0	0	0	0	3500	9000	2400	0	14,900	29.8	15
UR	0	0	0	6500	0	0	5500	0	14,400	0	0	0	0	0	0	0	26,400	52.8	8
IW	0	0	0	6500	0	0	16,500	5200	0	0	0	0	0	0	0	0	28,200	56.4	7
TFH	0	0	0	0	0	0	5500	20,800	0	0	0	0	0	0	0	0	26,300	52.6	9
TFL	0	0	0	0	0	0	0	0	0	9000	4200	0	7000	0	0	0	20,200	40.4	11
SSDI	0	0	21,000	0	12,200	0	0	0	0	0	0	0	0	0	0	0	33,200	66.4	3
CCP	0	0	0	19,500	0	0	0	0	9600	0	0	0	0	0	0	0	29,100	58.2	5
CLA	0	0	0	0	0	0	0	0	0	4500	8400	7800	0	0	0	0	20,700	41.4	10
CFI	0	0	0	0	0	0	0	0	0	4500	0	7800	7000	0	0	0	19,300	38.6	12
CF	43,000	0	0	0	0	0	0	0	0	0	0	0	0	0	0	0	43,000	86	1
DGW	0	38,000	0	0	0	0	0	0	0	0	0	0	0	0	0	0	38,000	76	2

Table 6.25 Factors influencing farmers perception of climate change—Garrett ranking: Chhattisgarh

Factors	Rank																Total
	1	2	3	4	5	6	7	8	9	10	11	12	13	14	15	16	
Drought	70	0	0	0	0	0	0	0	0	0	0	0	0	0	0	0	70
Hailstorm	0	0	0	0	0	0	0	0	0	0	70	0	0	0	0	0	70
Flood	0	0	0	0	0	0	0	0	0	0	0	70	0	0	0	0	70
AD	0	0	0	0	0	0	0	0	0	0	0	0	70	0	0	0	70
IPD	0	0	0	70	0	0	0	0	0	0	0	0	0	0	0	0	70
IFM	0	0	0	0	0	0	0	0	0	0	0	0	0	70	0	0	70
UR	0	70	0	0	0	0	0	0	0	0	0	0	0	0	0	0	70
IW	0	0	70	0	0	0	0	0	0	0	0	0	0	0	0	0	70
TFH	0	0	0	0	70	0	0	0	0	0	0	0	0	0	0	0	70
TFL	0	0	0	0	0	70	0	0	0	0	0	0	0	0	0	0	70
SSDI	0	0	0	0	0	0	70	0	0	0	0	0	0	0	0	0	70
CCP	0	0	0	0	0	0	0	0	0	0	0	0	0	0	70	0	70
CLA	0	0	0	0	0	0	0	0	0	0	0	0	0	0	0	70	70
CFI	0	0	0	0	0	0	0	0	0	70	0	0	0	0	0	0	70
CF	0	0	0	0	0	0	0	70	0	0	0	0	0	0	0	0	70
DGW	0	0	0	0	0	0	0	0	70	0	0	0	0	0	0	0	70
Total	70	70	70	70	70	70	70	70	70	70	70	70	70	70	70	70	1120
Garret-score	86	76	70	65	61	58	55	52	48	45	42	39	35	30	24	14	

Table 6.26 Rankings of factors representing climate change using Garrett ranking: Chhattisgarh

Factors	Rank																Total	Average	Rank
	1	2	3	4	5	6	7	8	9	10	11	12	13	14	15	16			
	Total score																		
Drought	6020	0	0	0	0	0	0	0	0	0	0	0	0	0	0	0	6020	86	1
Hailstorm	0	0	0	0	0	0	0	0	0	0	2940	0	0	0	0	0	2940	42	11
Flood	0	0	0	0	0	0	0	0	0	0	0	2730	0	0	0	0	2730	39	12
AD	0	0	0	0	0	0	0	0	0	0	0	0	2450	0	0	0	2450	35	13
IPD	0	0	0	4550	0	0	0	0	0	0	0	0	0	0	0	0	4550	65	4
IFM	0	0	0	0	0	0	0	0	0	0	0	0	0	2100	0	0	2100	30	14
UR	0	5320	0	0	0	0	0	0	0	0	0	0	0	0	0	0	5320	76	2
IW	0	0	4900	0	0	0	0	0	0	0	0	0	0	0	0	0	4900	70	3
TFH	0	0	0	0	4270	0	0	0	0	0	0	0	0	0	0	0	4270	61	5
TFL	0	0	0	0	0	4060	0	0	0	0	0	0	0	0	0	0	4060	58	6
SSDI	0	0	0	0	0	0	3850	0	0	0	0	0	0	0	0	0	3850	55	7
CCP	0	0	0	0	0	0	0	0	0	0	0	0	0	0	1680	0	1680	24	15
CLA	0	0	0	0	0	0	0	0	0	0	0	0	0	0	0	980	980	14	16
CFI	0	0	0	0	0	0	0	0	0	3150	0	0	0	0	0	0	3150	45	10
CF	0	0	0	0	0	0	0	3640	0	0	0	0	0	0	0	0	3640	52	8
DGW	0	0	0	0	0	0	0	0	3360	0	0	0	0	0	0	0	3360	48	9

Table 6.27 Factors influencing farmers' perception of climate change—Garrett ranking: West Bengal

Factors	Rank																Total
	1	2	3	4	5	6	7	8	9	10	11	12	13	14	15	16	
Drought	4	6	0	4	3	3	14	16	16	8	8	16	27	26	0	0	151
Hailstorm	0	0	0	36	55	36	24	0	0	0	0	0	0	0	0	0	151
Flood	0	0	0	0	0	0	0	0	0	0	0	0	0	0	151	0	151
AD	41	15	1	7	8	8	15	5	4	14	9	19	1	4	0	0	151
IPD	33	19	7	18	7	11	5	13	12	6	9	4	7	0	0	0	151
IFM	5	5	5	3	15	18	10	11	6	6	11	16	20	20	0	0	151
UR	29	31	46	11	9	11	11	0	1	0	0	2	0	0	0	0	151
IW	6	30	45	36	19	9	0	3	2	0	0	0	1	0	0	0	151
TFH	27	37	36	23	10	11	4	0	1	0	2	0	0	0	0	0	151
TFL	0	0	0	0	0	0	0	0	0	0	0	0	0	0	0	151	151
SSDI	2	5	0	8	10	19	15	19	21	10	25	13	4	0	0	0	151
CCP	0	0	0	3	1	1	3	6	5	13	13	26	39	41	0	0	151
CLA	1	0	2	1	3	7	11	19	10	26	26	24	16	5	0	0	151
CFI	0	0	0	2	1	4	7	14	18	8	14	22	29	32	0	0	151
CF	0	3	5	0	6	9	20	25	26	18	20	2	2	15	0	0	151
DGW	3	0	4	0	3	5	11	20	29	41	15	6	6	8	0	0	151
Total	151	151	151	152	150	152	150	151	151	150	152	150	152	151	151	151	2416
Garrett score	86	76	70	65	61	58	55	52	48	45	42	39	35	30	24	14	

Table 6.28 Rankings of factors representing climate change using Garrett ranking: West Bengal

Factors	Rank																Total	Average	Rank
	1	2	3	4	5	6	7	8	9	10	11	12	13	14	15	16			
	Total score																		
Drought	344	456	0	260	183	174	770	832	768	360	336	624	945	780	0	0	6832	45.25	12
Hailstorm	0	0	0	2340	3355	2088	1320	0	0	0	0	0	0	0	0	0	9103	60.28	6
Flood	0	0	0	0	0	0	0	0	0	0	0	0	0	0	3624	0	3624	24.00	15
AD	3526	1140	70	455	488	464	825	260	192	630	378	741	35	120	0	0	9324	61.75	5
IPD	2838	1444	490	1170	427	638	275	676	576	270	378	156	245	0	0	0	9583	63.46	4
IFM	430	380	350	195	915	1044	550	572	288	270	462	624	700	600	0	0	7380	48.87	8
UR	2494	2356	3220	715	549	638	605	0	48	0	0	78	0	0	0	0	10703	70.88	2
IW	516	2280	3150	2340	1159	522	0	156	96	0	0	0	35	0	0	0	10254	67.91	3
TFH	2322	2812	2520	1495	610	638	220	0	48	0	84	0	0	0	0	0	10749	71.19	1
TFL	0	0	0	0	0	0	0	0	0	0	0	0	0	0	0	2114	2114	14.00	16
SSDI	172	380	0	520	610	1102	825	988	1008	450	1050	507	140	0	0	0	7752	51.34	7
CCP	0	0	0	195	61	58	165	312	240	585	546	1014	1365	1230	0	0	5771	38.22	14
CLA	86	0	140	65	183	406	605	988	480	1170	1092	936	560	150	0	0	6861	45.44	11
CFI	0	0	0	130	61	232	385	728	864	360	588	858	1015	960	0	0	6181	40.93	13
CF	0	228	350	0	366	522	1100	1300	1248	810	840	78	70	450	0	0	7362	48.75	9
DGW	258	0	280	0	183	290	605	1040	1392	1845	630	234	210	240	0	0	7207	47.73	10

Chapter 7
Climate Change Impacts on Rice Yield in Western Region of India

K. Palanisami, Krishna Reddy Kakumanu, Udaya Sekhar Nagothu and C. R. Ranganathan

7.1 Introduction

The western region for the present study covers the Gujarat state. As stated in the methodology chapter, the present study is based on (i) primary data: farm survey among 103 farmers belonging to Navsari district and (ii) secondary data: time series data (1980–2010) on productivity (yield) of rice and climate variables from the district. The location of the study region is given in Fig. 7.1.

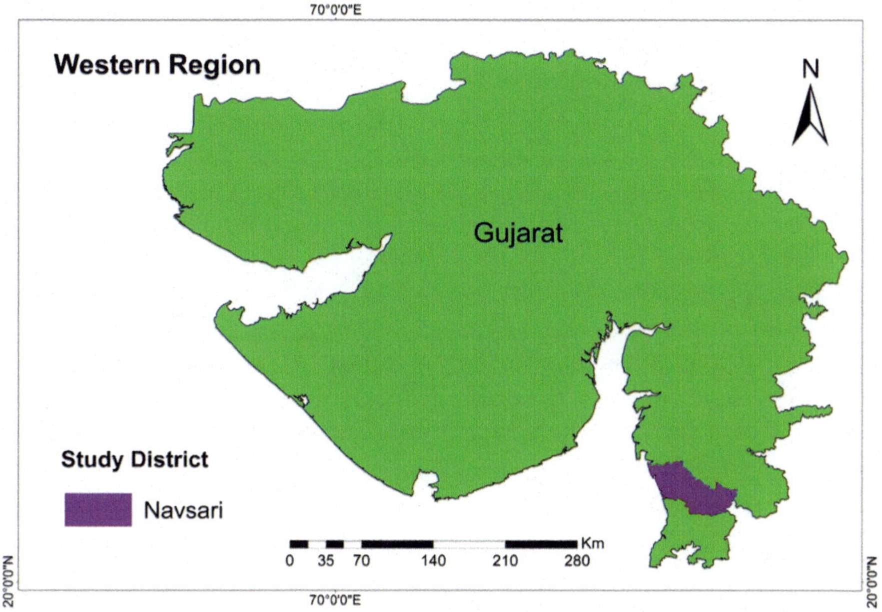

Fig. 7.1 Study region

K. Palanisami et al., *Climate Change and Future Rice Production in India*, India Studies in Business and Economics,
https://doi.org/10.1007/978-981-13-8363-2_7

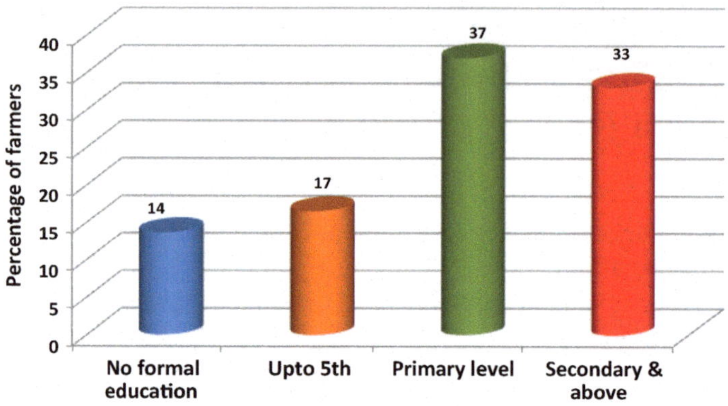

Fig. 7.2 Educational status of surveyed farmers—western region

7.2 Socio-economic Profiles of the Farmers

From the 103 farmers surveyed 99 were males and 4 were females. Out of the 103 farmers, 82 farmers have dry-land-based farming. The average age of the farmers is 49 years. About 37% of the farmers have primary-level education, 33% have studied secondary level and above, 17% have studied up to 5th and the rest, 14% have no formal education (Fig. 7.2).

7.3 Farm Information

Farmers in the study area have owned land and wells for irrigation where drip, sprinkler and other conventional types of irrigation methods are used. Farmers in Gujarat derive income from rice during kharif season and from other crops during all the three seasons.

(a) **Physical assets**

(i) **Land owned**: Among the sample farmers, 98 farmers owned irrigated lands and 38 farmers owned both irrigated and dry lands. Table 7.1 gives the summary statistics of the type of land owned by the farmers.

Table 7.1 Summary statistics of land owned (hectares)

Land type	Mean	S.D.	Min.	Max.	Count
Irrigated	0.96	0.74	0.20	3.64	98
Dry	0.73	0.83	0.20	4.86	38
Overall	0.90	0.77	0.20	4.86	136

Table 7.2 Distribution of number of wells owned

No. of wells	Number of farmers	
	Open wells	Borewells
0	91	38
1	12	61
2	0	3
3	0	1

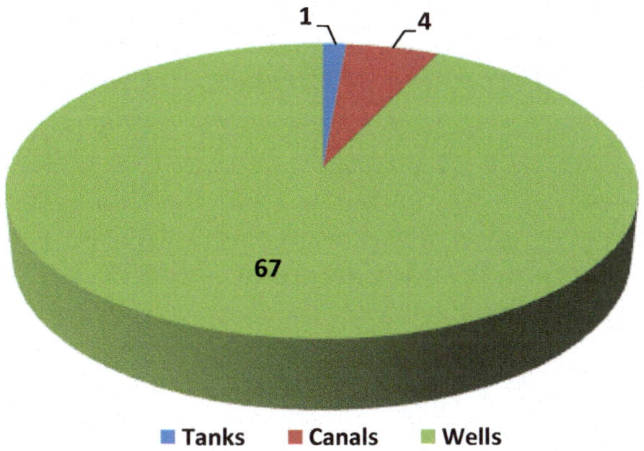

Fig. 7.3 Frequency distribution of sources of irrigation

The mean area under irrigation is 0.96 and 0.73 ha under dry lands, and the average land area is 0.90 ha. Irrigated landholding has a standard deviation of 0.74 ha, and the minimum and maximum size of the overall landholdings are, respectively, 0.20 and 4.86 ha.

(ii) **Wells owned**: Borewells are most common in the study area. Table 7.2 gives the frequency distribution of number of borewells owned by the farmers.
About 91% of the farmers do not have open wells and about 38% do not have borewells. Further, 61 farmers (59%) have one borewell each and 12% farmers possess one open well.

(iii) **Sources of irrigation**: Farmers have three sources of irrigation, viz. tank, canal and well.
Well is the main source of irrigation for 67 farmers (65%) as shown in Fig. 7.3.

(iv) **Investment in irrigation**: Farmers' investment is related to digging wells and installation of the microirrigation systems.

(v) **Area under rice**: From the total 103 farmers, 96 of them have cultivated rice during the kharif season. During this season, on the average, every farmer cultivates rice in 0.75 ha. Other crops are grown covering an area of 0.8, 0.75, 0.57 ha in kharif, rabi and summer seasons, respectively.

7.4 Climate Change Events, Shocks and Their Effects as Perceived by Farmers

With a view to have an in-depth understanding of farmers perception on the climate change, its effects on agriculture and various adaptation strategies/techniques followed by the farmers, detailed survey was conducted and the results are discussed below.

7.4.1 Perception of Farmers on Climate Change Events

Farmers were asked to rank the climate change events as listed in the methodology chapter according to their perceptions of climate change. To find what the farmers perceive as the most significant effects of climate change, Garrett Ranking technique was applied. The results are given in Tables 7.3 and 7.4. Table 7.4 shows that irregular weather is the most important climate change event (Rank 1) perceived by rice farmers in western region. It earned an average Garrett Score of 79.34. As given in Table 7.3, it was given Ranks 1, 2 and 3 by 61, 25 and 12 farmers, respectively. The second and third important climate change events are, respectively, high temperature fluctuation and untimely rain.

Friedman Rank Test was applied to test for differences between rankings of the climate change events (listed in Table 7.4) by the farmers. The methodology as described in Chap. 4 was followed. The calculated value of Friedman's F statistics was 884.58 while the table value (corresponding to 14 degrees of freedom at 5% level of significance) is 23.68. This shows that the farmers did not consistently rank the events; thus, the farmers' perceptions of climate change effects differ across the sample chosen. The same conclusion was arrived at by applying Kendall's coefficient of concordance also.

7.4.2 Effects of Shocks

Farmers were questioned on their perception about the effects of climatic shocks. They were asked to rank seven effects of climatic shocks, as described in the methodology chapter. But, they ranked only five climatic effect shocks. Farmers ranked loss of income as the most important shock (Tables 7.5 and 7.6). Out of 103 farmers, 95 of them ranked loss of income as the most important shock. The second and third ranks are given to the decline in health and food insecurity/shortage. This is understandable because the decline in income will reflect in lesser spending on consumption leading to decline in health and food insecurity.

Table 7.3 Factors influencing farmers perception of climate change—Garrett ranking: western region

Factors	Rank															Total
	1	2	3	4	5	6	7	8	9	10	11	12	13	14	15	
Hailstorm	0	0	0	0	2	0	0	0	0	0	0	0	2	24	75	103
Flood	0	0	1	9	7	2	0	0	0	0	0	3	38	39	4	103
AD	0	0	2	1	9	6	12	13	7	11	7	18	14	3	0	103
IPD	0	5	2	14	20	22	17	12	8	1	2	0	0	0	0	103
IFM	0	2	3	2	7	2	4	13	13	10	25	14	4	3	1	103
UR	10	24	29	8	6	8	4	6	3	1	3	0	1	0	0	103
IW	61	25	12	2	1	0	1	0	0	1	0	0	0	0	0	103
TFH	26	33	33	8	0	0	0	0	1	0	0	0	1	1	0	103
TFL	4	4	8	8	5	2	4	1	2	2	1	3	16	25	18	103
SSDI	0	1	0	19	10	33	11	11	8	4	3	1	0	0	2	103
CCP	0	2	4	7	8	7	10	9	19	20	9	8	0	0	0	103
CLA	0	0	0	1	2	6	6	5	5	13	26	26	10	2	1	103
CFI	0	1	4	1	1	1	7	8	15	9	17	24	10	5	0	103
CF	0	0	0	2	4	5	21	14	16	26	8	5	1	1	0	103
DGW	2	6	5	21	21	9	6	11	6	5	2	1	6	0	2	103
Total	103	103	103	103	103	103	103	103	103	103	103	103	103	103	103	1545
Garrett score	85	75	69	64	60	56	53	50	47	43	40	36	31	25	15	

Table 7.4 Rankings of factors representing climate change using Garrett ranking: western region

Factors	Rank															Total score	Average score	Rank
	1	2	3	4	5	6	7	8	9	10	11	12	13	14	15			
	Total score																	
Hailstorm	0	0	0	0	120	0	0	0	0	0	0	0	62	600	1125	1907	18.51	15
Flood	0	0	69	576	420	112	0	0	0	0	0	108	1178	975	60	3498	33.96	14
AD	0	0	138	64	540	336	636	650	329	473	280	648	434	75	0	4603	44.69	10
IPD	0	375	138	896	1200	1232	901	600	376	43	80	0	0	0	0	5841	56.71	4
IFM	0	150	207	128	420	112	212	650	611	430	1000	504	124	75	15	4638	45.03	9
UR	850	1800	2001	512	360	448	212	300	141	43	120	0	31	0	0	6818	66.19	3
IW	5185	1875	828	128	60	0	53	0	0	43	0	0	0	0	0	8172	79.34	1
TFH	2210	2475	2277	512	0	0	0	0	47	0	0	0	31	25	0	7577	73.56	2
TFL	340	300	552	512	300	112	212	50	94	86	40	108	496	625	270	4097	39.78	13
SSDI	0	75	0	1216	600	1848	583	550	376	172	120	36	0	0	30	5606	54.43	6
CCP	0	150	276	448	480	392	530	450	893	860	360	288	0	0	0	5127	49.78	7
CLA	0	0	0	64	120	336	318	250	235	559	1040	936	310	50	15	4233	41.10	12
CFI	0	75	276	64	60	56	371	400	705	387	680	864	310	125	0	4373	42.46	11
CF	0	0	0	128	240	280	1113	700	752	1118	320	180	31	25	0	4887	47.45	8
DGW	170	450	345	1344	1260	504	318	550	282	215	80	36	186	0	30	5770	56.02	5

Table 7.5 Rankings of effects of climatic shocks by farmers

Factors	Rank					Total
	1	2	3	4	5	
LI	95	8	0	0	0	103
FIS	4	26	14	32	27	103
DL	1	16	35	30	21	103
DC	2	20	32	21	28	103
DH	1	33	22	20	27	103
Total	103	103	103	103	103	515
Garrett score	75	60	50	40	25	

Table 7.6 Garrett ranking of effects of climatic shocks

Factors	Rank					Total score	Average score	Rank
	1	2	3	4	5			
	Total score							
LI	7125	480	0	0	0	7605	73.8	1
FIS	300	1560	700	1280	675	4515	43.8	3
DL	75	960	1750	1200	525	4510	43.8	4
DC	150	1200	1600	840	700	4490	43.6	5
DH	75	1980	1100	800	675	4630	45.0	2

7.4.3 Opinions of Farmers on Government Initiatives

From time to time, both Central and State governments implement several mitigations and long-term measures to offset the negative effects of climate change on agriculture. With a view to know the awareness of these measures among the farmers, they were questioned on six government initiatives (as described in the methodology chapter). They were asked to state their replies as either 'yes', or 'no' or 'not aware'. Figure 7.4 gives a pictorial representation of the awareness of these initiatives. About 91% of the farmers are well aware of subsidized inputs (like fertilizers, drip irrigation and machinery), 57% know about field demos on the adaptation practices. Farmers have no knowledge about all other initiatives like weather-based crop insurance schemes and establishment of village knowledge centres. To sum up, farmers had poor knowledge of government initiatives, and so suitable mechanisms should be formed by the government to popularize government initiatives these two initiatives. Table 7.7 provides a summary of the opinions of farmers on government initiatives.

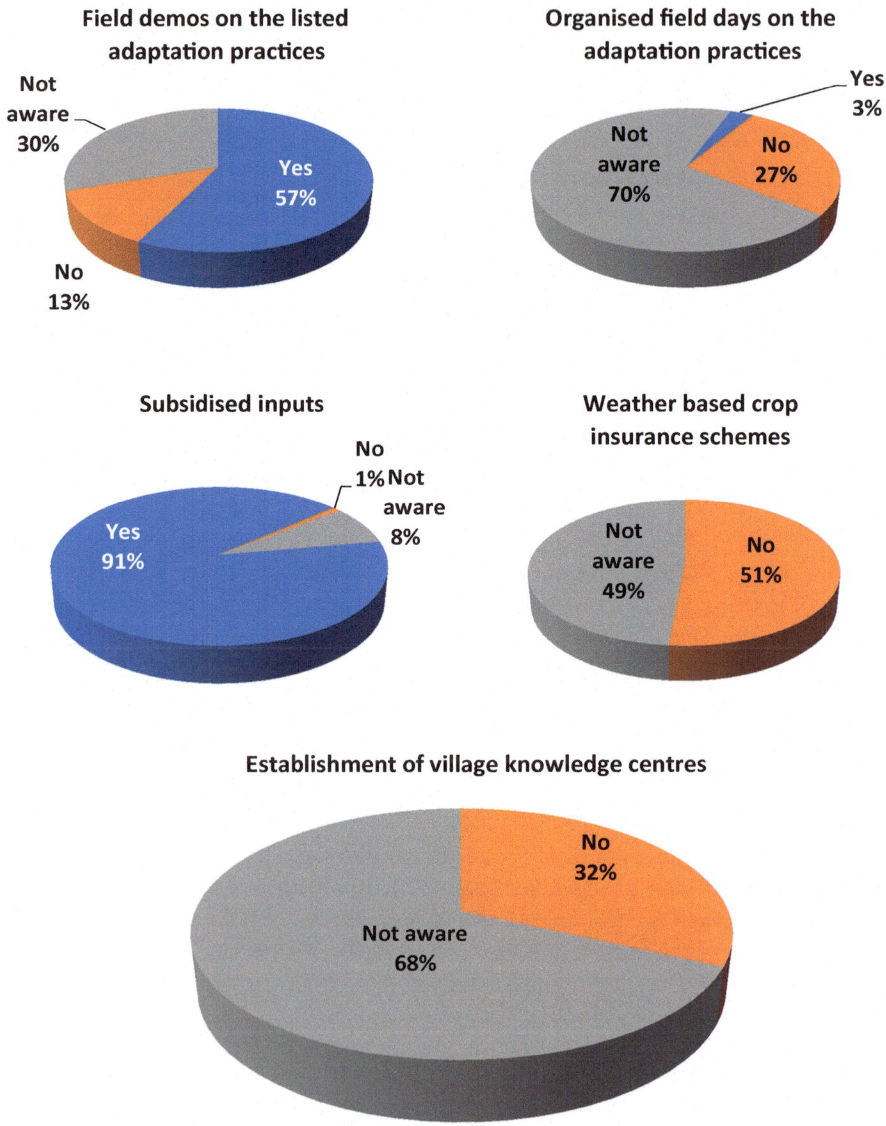

Fig. 7.4 Opinions of farmers of western region on various government initiatives

Table 7.7 Farmers' opinions on government initiatives

Govt. initiatives	Yes	No	Not aware
Field demos on the listed adaptation practices	59	13	31
Organized field days on the adaptation practices	3	28	72
Subsidized inputs (fertilizer, drip and machinery)	94	1	8
Weather-based crop insurance schemes	0	53	50
Establishment of village knowledge centres, climate cell, etc.	0	33	70

7.5 Analysis of the Effect of Climate Change on Rice Yield

7.5.1 Data

One important objective of the study is to estimate the effects of climate variables on the yield of rice crop using historical data. Hence, we used a time series data set of yield and climate variables pertaining to 31 years from 1980 to 2010. The data on temperature corresponds to the rice growing season and the time series data on rainfall was collected for the entire year to account for rainfall falling directly on the crop and interseasonal water accumulation within a year (Iski and Devadoss 2006; Kim and Pang 2009). The summary statistics of all the variables are discussed below:

(a) **Yield (tons per ha)**

The yield data (Table 7.8; Fig. 7.5) shows a general upward trend over the years. Trend equation fitted to yield data shows that the coefficient of trend is strongly significant (1% level).

(b) **Maximum Temperature** (°C)

The maximum temperature from 1980 to 2010 had an overall average of 32.08 °C with a standard deviation of 0.546 °C. It shows a slightly increasing trend over the years (Fig. 7.6) (Table 7.9).

(c) **Minimum Temperature** (°C)

The minimum temperature had an average of 20.83 °C with a standard deviation of 1.11 °C. Compared with maximum temperature, it had greater variability. Further, the minimum temperature had a clear increasing trend over the years (Fig. 7.7) (Table 7.10).

(d) **Rainfall** (mm)

During the study period, 1980–2010, the rainfall in the region had an average of 1593 mm with a standard deviation of 472 mm. It had a wide range between 961 and 2439 mm. Further, the rainfall did not show any trend over the years (Fig. 7.8) (Table 7.11).

Table 7.8 Summary statistics on rice yield and trend equation (tons/ha)

Mean	Std.	Min.	Max.	Trend equation	R^2
1.46	0.31	0.81	1.96	$Y = 1.12 + 0.0227^{***}$	0.45

*** denotes significance at 1% level

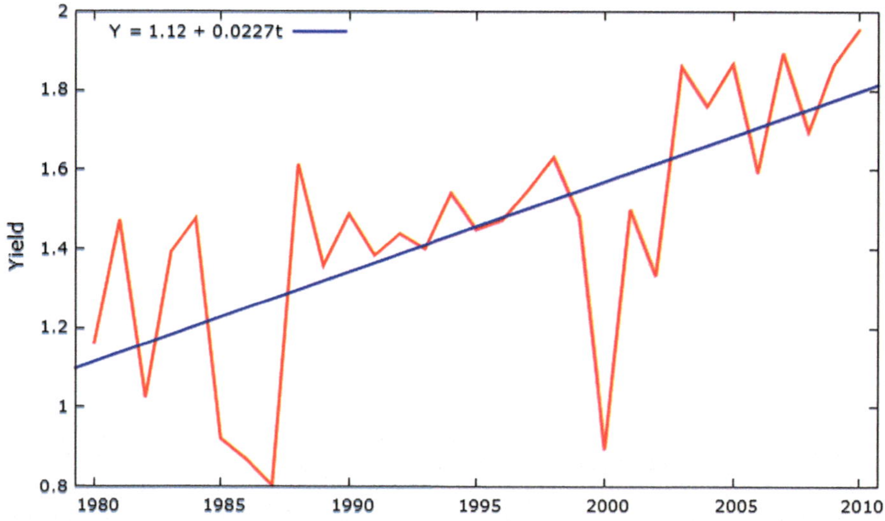

Fig. 7.5 Trend in yield of rice per ha in Gujarat

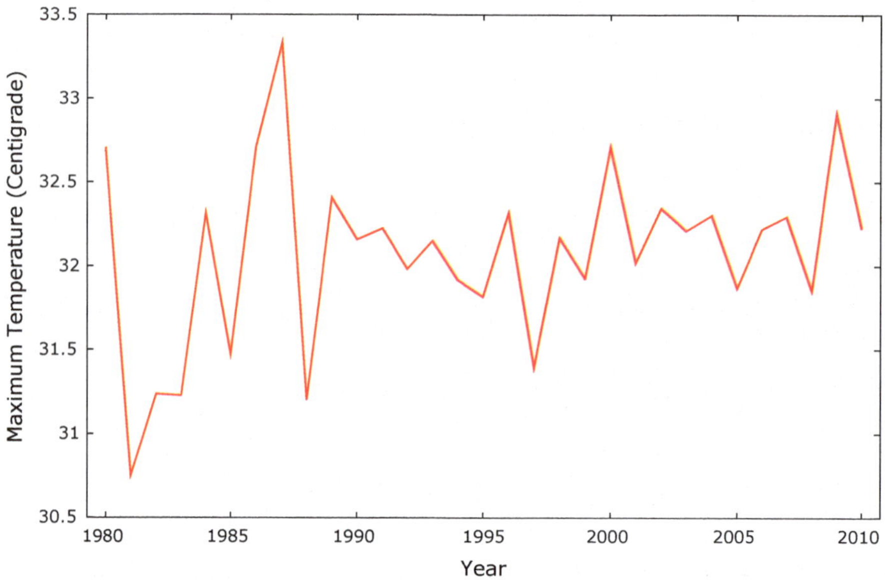

Fig. 7.6 Trend in maximum temperature in Gujarat

Table 7.9 Descriptive statistics of maximum temperature in Gujarat (°C)

Mean	Minimum	Maximum	Std. dev.
32.08	30.76	33.34	0.546

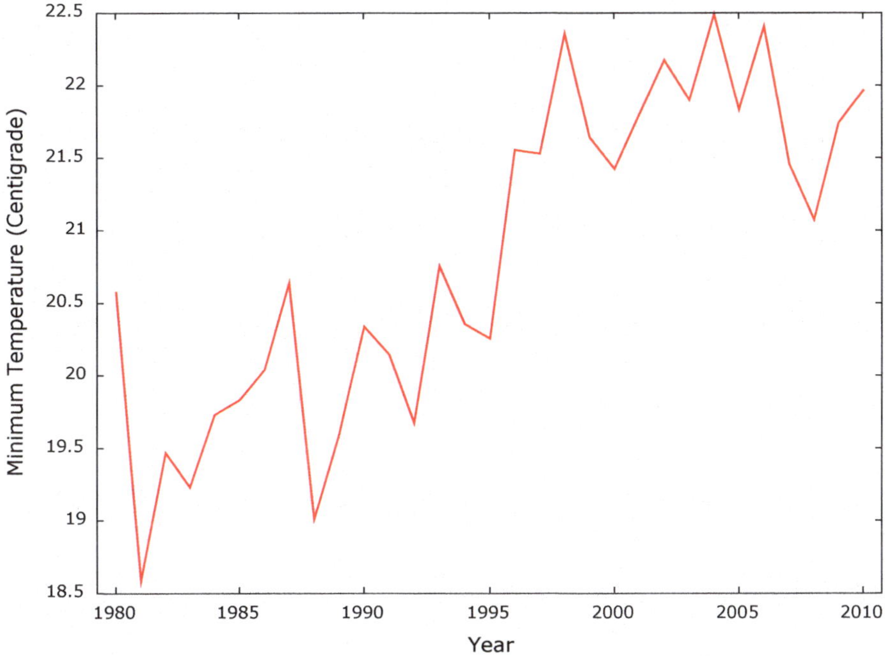

Fig. 7.7 Trend in minimum temperature in Gujarat

Table 7.10 Descriptive statistics of minimum temperature in Gujarat (°C)

Mean	Minimum	Maximum	Std. dev.
20.83	18.59	22.49	1.11

Table 7.11 Descriptive statistics of rainfall (mm) in Gujarat

Mean	Minimum	Maximum	Std. dev.
1593	961	2439	472

7.5.2 Test of Normality of Yield Data

The normality of yield data was tested using Shapiro–Wilk W test and Lilliefors test. The tests showed that the null hypothesis that the data is normal cannot be rejected. The computed value of Shapiro–Wilk W test was 0.9316 with a p-value of 0.0584. Since this value is greater than 0.05, the null hypothesis was not rejected. Similarly, the test statistic for Lilliefors test was 1.506 with a p-value of 0.07 which is also greater than 0.05. So, both tests did not reject the null hypothesis that yield data is normal.

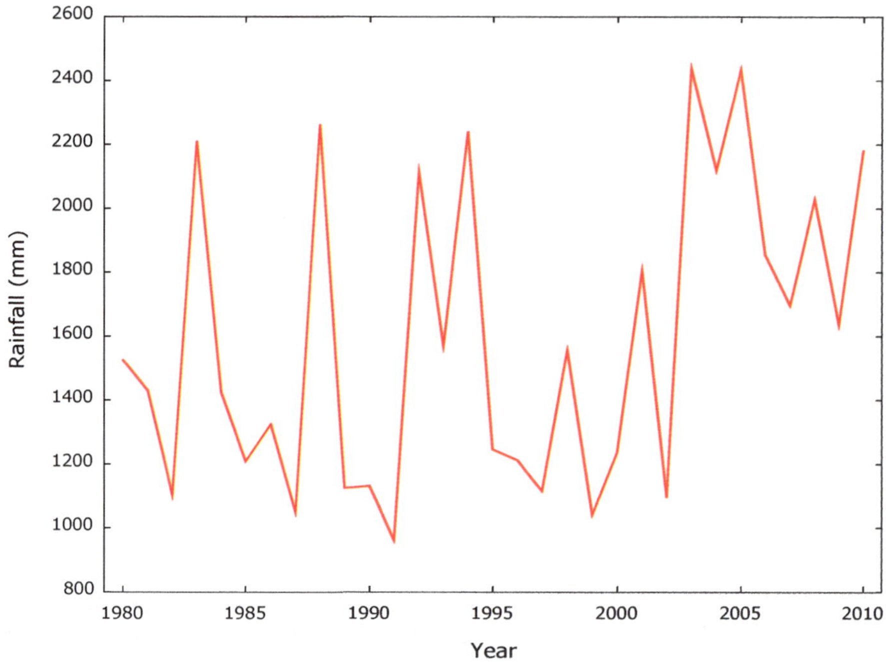

Fig. 7.8 Trend in rainfall in Gujarat

7.5.3 *Estimation of Just–Pope Production Function*

As discussed in the methodology chapter, to account for both mean and variance in yield, Just-Pope production function (Just and Pope 1978, 1979) was employed. The dependent variable was yield, and independent variables were maximum and minimum temperatures and rainfall (introduction of square and interaction terms of maximum temperature, minimum temperature and rainfall resulted in convergence problems for estimation and so they were omitted). Table 7.12 gives the results of the estimated empirical models.

Table 7.12 shows that trend and all the climate variables are strongly significant for the mean function implying that climate change effects are important. For the variance function, minimum temperature and rainfall have significant negative effects and so they are risk-decreasing variables. That is, when the values of these variables are increased, variability in yield decreases.

The elasticities for all climatic variables at mean levels are presented in Table 7.13 for both mean function and variance function. The elasticity for the maximum temperature is −1.8082 implying that 1% increase in maximum temperature will induce a decrease of 1.81% in mean yield. Similarly, the elasticity coefficient for minimum temperature is 0.5978 and so 1% increase in minimum temperature will increase the mean yield by a small percentage of 0.6%. The elasticity coefficient for rainfall is

Table 7.12 Estimation results of Just–Pope production function

	Coefficient	S. error	p-value
Mean function			
Constant (β_0)	2.5794	1.8944	0.173
Trend (β_1)	0.0146**	0.0058	0.011
Max$T$ (β_2)	−0.0827**	0.0399	0.038
Min$T$ (β_3)	0.0429**	0.0196	0.0287
Rain (β_4)	0.0003***	0.0001	0
Variance function			
Constant	−1.7338	19.3450	0.929
Trend	0.1356	0.0922	0.142
Max$T$	0.8925	0.5541	0.107
Min$T$	−1.4207*	0.7386	0.054
Rain	−0.0025***	0.0009	0.004
Model statistics			
Log-likelihood	16.931		
AIC	−13.862		

*, **, *** denotes significance at 10%, 5% and 1% level

Table 7.13 Elasticities of climatic variables at means

Yield function	Climate variable	Elasticity
Mean	Maximum temperature	−1.8082
	Minimum temperature	0.5978
	Rainfall	0.2797
Variance	Maximum temperature	28.4503
	Minimum temperature	−28.9105
	Rainfall	−3.9189

0.2797 which is very small and so 1% increase in rainfall will induce a small increase of 0.3% in mean yield. When all the three variables are simultaneously increased by 1%, we can expect a small decrease of 0.93% in yield.

For the variance function, the elasticity coefficient of maximum temperature is −28.4503 implying an increase in variability in yield of 28.5% for 1% increase in maximum temperature. Similar meanings can be given for the elasticities of minimum temperature and rainfall. All together, we can expect a decrease of 4.4% in yield variability when all the three variables are increased by 1%. So, we conclude that all the three variables put together are risk decreasing in this sense.

Table 7.14 Climate change impacts on percentage of mean yield and yield variability

Scenario	Period	Percentage change in				
		T_{max}	T_{min}	Rainfall	Mean yield	Yield variability
RCP4.5	2021–2050	5.83	9.65	23.72	**1.9**	−206.0
	2071–2100	8.15	15.40	4.58	**−5.8**	−224.3
RCP8.5	2021–2050	5.44	10.26	38.57	5.6	−285.4
	2071–2100	15.26	27.11	98.82	14.7	−729.9

7.5.4 Climate Change Scenarios and Rice Yield Changes

The estimated elasticities given in Table 7.13 were used to estimate the impacts of climate change on mean yield and yield variability of yield. For this purpose, two climate change scenarios, viz. RCP4.5 and RCP8.5 (as discussed in Chap. 3) were considered and the results are given in Table 7.14.

The table shows that there will be an increase in percentages of all the three climatic variables during 2021–2050 and 2071–2100 under both scenarios RCP4.5 and RCP8.5. There will be an increase of 5–6% in maximum temperature in the first period. During the second period, the increase will be between 8 and 15%. The predicted increase in minimum temperature is higher. It varies from about 10% for the first period to 27% for the second period. RCP4.5 scenario predicts an increase of 24% in rainfall during the first period, whereas it predicts only 5% increase for the second period. The RCP8.5 scenario predicts very high increase in rainfall varying between 39 and 99% in the two periods.

The table shows that as per RCP4.5 and RCP8.5 scenarios, there will be a marginal increase in percentages of mean rice yield, viz. 1.9 and 5.6% during the period 2021–2050. Further, during this period, there will a drastic reduction in variability of yield varying from 206 to 285%. For the period 2071–2100, according to RCP4.5 scenario, there will be a reduction in mean yield of 5.8% while RCP8.5 scenario predicts an increase in mean yield of 14.7%. There will be a reduction of 224 and 730% variability in yield as per the two scenarios during 2071–2100.

7.6 Summary

Irregular weather is the most important climate change event (Rank 1) perceived by rice farmers in western region, followed by temperature fluctuation and untimely rain. Farmers ranked loss of income as the most important shock, followed by a loss of assets.

Farmers in the region have very good awareness of subsidized inputs and field demos on the adaptation practices. But, they do not have knowledge about all other initiatives like weather-based crop insurance schemes and establishment of village knowledge centres, thus, warranting suitable measures by the government to popularize government schemes.

Based on historical data, the yield of rice shows a marginal upward trend in the region over the years, although there was a decline in yield in some years. Both maximum and minimum temperatures exhibited an increasing trend over the years. Annual rainfall had a wide range with no trend over the years. Maximum temperature has a significant negative effect on rice yield, whereas minimum temperature and rainfall have significant negative effects. There is no nonlinear effect of the climate variables on rice yield.

The two climate change scenarios, viz. RCP4.5 and RCP8.5 predict a marginal increase in percentages of mean rice yield during the period 2021–50. But during this period, there will a drastic reduction in variability of yield. For the period 2071–2100, the two scenarios predict inconsistent changes in mean yield.

Chapter 8
Climate Change Impacts on Rice Yield in Central Region of India

K. Palanisami, Krishna Reddy Kakumanu, Udaya Sekhar Nagothu and C. R. Ranganathan

8.1 Introduction

The Central Region for the present study covers two states: Madhya Pradesh and Maharashtra. The districts selected from the two States were: Balaghat and Bhandara respectively. As stated in the methodology chapter, the present study is based on (i) primary data: farm survey among 70 farmers from Madhya Pradesh and 100 farmers from Maharashtra of the two districts and (ii) Secondary data: time series data (1980–2010) on productivity (yield) of rice and climate variables from the two districts. Figure 8.1 gives the map of the study area.

8.2 Socio-economic Profiles of the Farmers

Among the 170 farmers surveyed, 166 were males and 4 were females. All the 70 farmers in Madhya Pradesh, have dry land-based crop typology while in Maharashtra, 31 farmers have irrigated crop based farming, 64 farmers have irrigated crop with supplemental irrigation and 5 farmers have dry land irrigation. The average age of the farmers in the two States are respectively, 46 and 51 years and farmers in Maharashtra are relatively older than those in Madhya Pradesh. In general, about 6% of the farmers have no formal education, 21% have studied up to 5th, 55% have studied up to primary level and the rest, 18% have studied up to secondary level and above (Fig. 8.2). About 59% of the farmers in Maharashtra are having education up to primary level whereas about 43% farmers in Madhya Pradesh have education level of secondary and above.

K. Palanisami et al., *Climate Change and Future Rice Production in India*, India Studies in Business and Economics,
https://doi.org/10.1007/978-981-13-8363-2_8

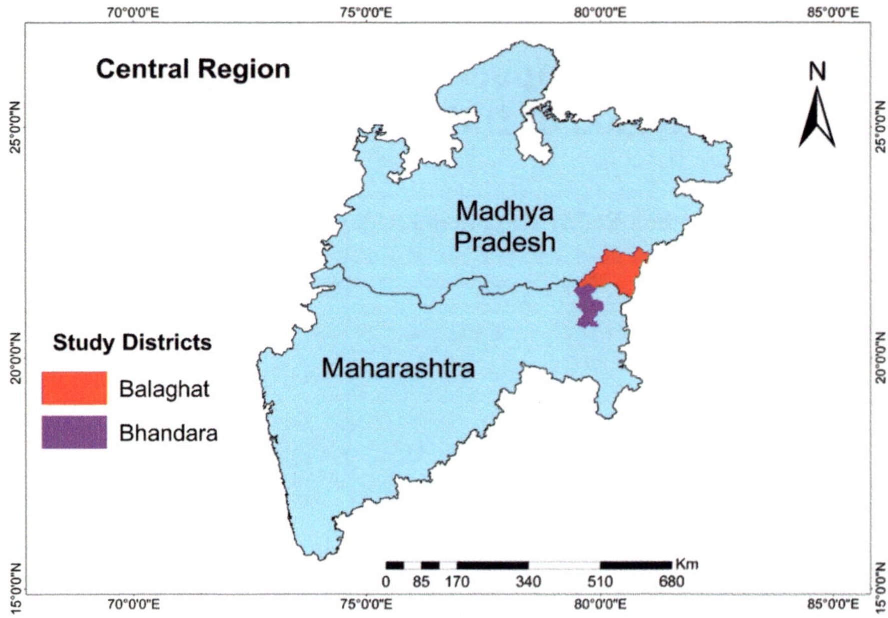

Fig. 8.1 Study regions

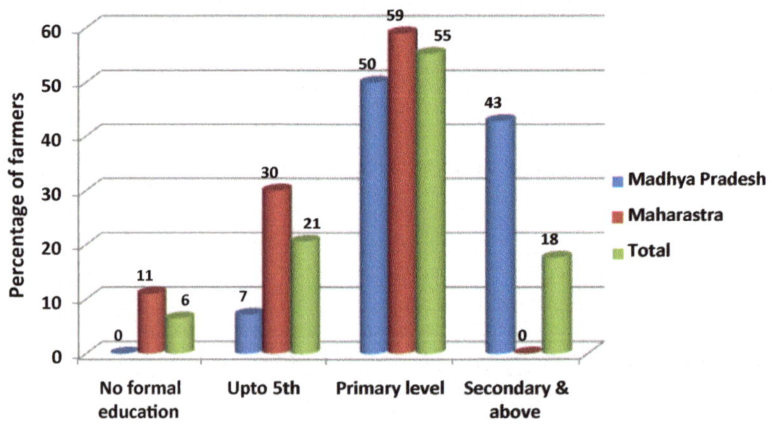

Fig. 8.2 Educational status of surveyed farmers

8.3 Farm Information

Farmers in the study region had irrigated lands supported by wells as the major source of irrigation. In addition to their traditional gravity based irrigation, they also used drip and sprinkler systems. Most of the farmers in these states derive income from rice and other crops mainly during Kharif season although few are raising other crops during Rabi season.

(a) **Physical assets**:

(i) **Land owned**: All the farmers own irrigated lands. Table 8.1 gives the summary statistics of the land size.

The mean area under irrigation varies from 1.41 to 3.37 ha across the two states and the overall mean is 2.22 ha. Land holdings in Madhya Pradesh has relatively higher standard deviation (2.30 ha) and the minimum and maximum holdings are respectively 0.81 and 16.19 ha in Madhya Pradesh and 0.20 and 7.28 ha in Maharashtra.

(ii) **Wells owned**: Farmers in Madhya Pradesh possess only open wells whereas those from Maharashtra possess both open and bore-wells. Table 8.2 gives the frequency distribution of number of wells owned by the farmers.

Out of 70 farmers in Madhya Pradesh, only 22 (31%) own one or two open wells and the remaining 69% don't own any well at all. About 42% of the farmers in Maharashtra possess, either open or bore well. Overall about 75% of the farmers in the region don't own open wells and 46% don't own bore wells.

Table 8.1 Summary statistics of irrigated land owned (hectares)

State	Mean	S.D.	Min	Max
Madhya Pradesh	3.37	2.30	0.81	16.19
Maharashtra	1.41	1.42	0.20	7.28
Overall	2.22	2.07	0.20	16.19

Table 8.2 Distribution of number of wells owned by farmers

No. of wells	Number of farmers				
	Madhya Pradesh	Maharashtra		Total	
	Open wells	Open wells	Bore wells	Open wells	Bore wells
0	48	79	79	127	79
1	17	15	19	32	19
2	5	5	1	10	1
3	–	–	1	–	1
>3	–	1	–	1	–

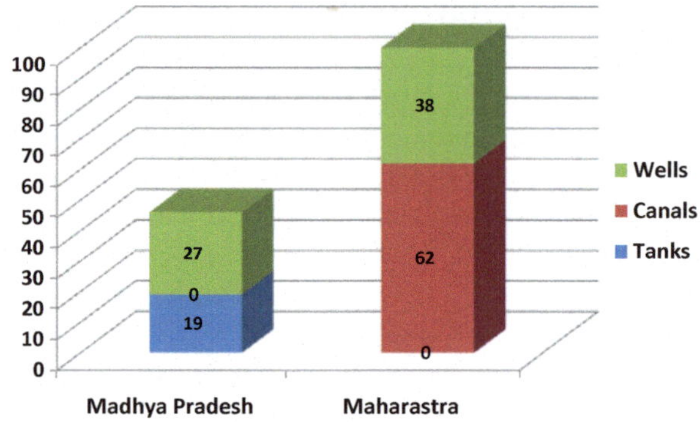

Fig. 8.3 Frequency distribution of sources of irrigation

Table 8.3 Rice area (ha) cultivated by farmers in the region

States	Rice			Other crops		
	Kharif	Rabi	Summer	Kharif	Rabi	Summer
Madhya Pradesh	3.37 (70)	0 (0)	0 (0)	0 (0)	1.33 (27)	0 (0)
Maharashtra	1.21 (81)	1.21 (2)	0.92 (13)	1.72 (20)	0 (0)	0 (0)

Figures in brackets denote the number of farmers who raised the particular crop in the particular season

(iii) **Sources of irrigation**: Like in other regions, farmers in this region also benefitted from three sources of irrigation, viz., canals, wells and tanks. Farmers of Madhya Pradesh mostly use only wells and tanks whereas Maharashtra farmers mostly use wells and canals (Fig. 8.3).

(iv) **Investment in irrigation**: Field surveys in the selected districts in the states indicated that the main investment type is well investment followed by investment in micro irrigation systems like drip and sprinkler irrigation.

(v) **Area under rice**: All farmers in Madhya Pradesh have cultivated rice during Kharif season only and 81% of the farmers in Maharashtra have grown rice during this season. During this season, average area per farmer under rice in the two states are respectively 3.37 and 1.21 ha. In Madhya Pradesh, out of 70 farmers surveyed, 46 farmers have irrigated rice with an average area of 2.2 ha. During Rabi and summer seasons only very small number of farmers in Maharashtra have grown rice. Some farmers in Madhya Pradesh grow other crops during Rabi season. In Maharashtra other crops are grown only during Kharif season (Table 8.3).

8.4 Climate Change Events, Shocks and Their Effects as Perceived by Farmers

The detailed survey conducted in the region helped to get an idea of farmers perception on the climate change, its effects on agriculture and various adaptation strategies/techniques followed by the farmers and the details are discussed below.

8.4.1 Perception of Farmers on Climate Change Events

Farmers were asked to rank the 16 events as listed in the methodology chapter according to their perceptions of climate change. To find what the farmers perceive as the most significant effects of climate change, Garrett's ranking technique as described in Chap. 4 was used. The results are given in Tables 8.4 and 8.5. Table 8.5 shows that the drought is the most important climate change event (Rank 1) perceived by Rice farmers in Central Region with an average Garret Score of 68.1. As given in Table 8.4 it was given Rank 1 by 45 farmers and rank 2 by 35 farmers. The second and third important climate change events are increase in pest damage and untimely rains respectively. Similar analysis was done separately for the two states in the region and the results are presented in Appendix (Tables 8.18, 8.19, 8.20, 8.21).

Friedman Rank Test was applied to test for differences between rankings of the climate change events (listed in Table 8.4) by the farmers. The methodology as described in Chap. 4 was followed. The calculated value of Friedman's F statistics was 397.65 while the table value (corresponding to 15 degrees of freedom at 5% level of significance) is 24.996. This shows that the farmers did not consistently rank the events, thus, the farmers' perceptions of climate change effects differ across the sample chosen. The same conclusion was arrived at by applied Kendall's coefficient of concordance also. Similar analysis was done for individual states in the region. The results show that there is no consistency in the rankings of the events by the farmers.

8.4.2 Effects of Shocks

Farmers were questioned on their perception about the effects of climatic shocks. They were asked to rank 7 effects of climatic shocks, as described in methodology chapter. Farmers ranked loss of income as the most important shock (Tables 8.6 and 8.7). Out of 170 farmers, 61 of them have given rank 1 and 77 of them have given rank 2 to this effect. The second and third ranks are given to decline in crop yield and loss of assets respectively.

Table 8.4 Factors influencing farmers perception of climate change—Garrett ranking: Central Region

Factors	Rank																Total
	1	2	3	4	5	6	7	8	9	10	11	12	13	14	15	16	
Drought	45	35	20	12	12	4	1	1	1	38	1	0	0	0	0	0	170
Hailstorm	17	2	4	2	20	0	0	0	1	53	0	13	20	10	15	13	170
Flood	29	10	3	3	0	1	13	14	9	2	5	0	21	59	0	1	170
AD	0	9	25	14	2	2	13	12	20	6	0	11	7	42	0	7	170
IPD	20	65	20	3	14	2	6	7	20	4	0	2	4	0	1	2	170
IFM	9	5	26	23	10	1	0	0	0	6	1	13	13	35	13	15	170
UR	9	7	24	50	57	10	10	0	0	0	0	3	0	0	0	0	170
IW	5	1	13	19	9	4	4	2	90	12	10	1	0	0	0	0	170
TFH	0	0	0	0	1	105	26	10	10	4	4	7	0	0	0	3	170
TFL	0	0	5	6	7	2	105	22	8	4	8	0	1	1	0	1	170
SSDI	0	0	0	1	1	3	6	7	0	14	97	20	10	10	1	0	170
CCP	0	0	0	18	18	7	1	5	0	0	10	90	9	4	1	7	170
CLA	0	0	0	0	2	8	3	12	7	7	31	0	0	0	100	0	170
CFI	0	0	0	0	4	3	1	3	0	6	0	39	3	9	5	97	170
CF	0	0	1	1	0	14	8	6	3	0	7	10	104	6	9	1	170
DGW	0	0	0	1	10	8	6	111	26	7	0	0	0	0	0	1	170
Total	134	134	141	153	167	174	203	212	195	163	174	209	192	176	145	148	2720
Garret-score	86	76	70	65	61	58	55	52	48	45	42	39	35	30	24	14	

Table 8.5 Rankings of factors representing climate change using Garrett ranking: Central Region

Factors	Rank																		
	1	2	3	4	5	6	7	8	9	10	11	12	13	14	15	16	Total score	Average score	Rank
	Total score																		
Drought	3870	2660	1400	780	732	232	55	52	48	1710	42	0	0	0	0	0	11581	68.1	1
Hailstorm	1462	152	280	130	1220	0	0	0	48	2385	0	507	700	300	360	182	7726	45.4	11
Flood	2494	760	210	195	0	58	715	728	432	90	210	0	735	1770	0	14	8411	49.5	8
AD	0	684	1750	910	122	116	715	624	960	270	0	429	245	1260	0	98	8183	48.1	9
IPD	1720	4940	1400	195	854	116	330	364	960	180	0	78	140	0	24	28	11329	66.6	2
IFM	774	380	1820	1495	610	58	0	0	0	270	42	507	455	1050	312	210	7983	47.0	10
UR	774	532	1680	3250	3477	580	550	0	0	0	0	117	0	0	0	0	10960	64.5	3
IW	430	76	910	1235	549	232	220	104	4320	540	420	39	0	0	0	0	9075	53.4	6
TFH	0	0	0	0	61	6090	1430	520	480	180	168	273	0	0	0	42	9244	54.4	4
TFL	0	0	350	390	427	116	5775	1144	384	180	336	0	35	30	0	14	9181	54.0	5
SSDI	0	0	0	65	61	174	330	364	0	630	4074	780	350	300	24	0	7152	42.1	13
CCP	0	0	0	1170	1098	406	55	260	0	0	420	3510	315	120	24	98	7476	44.0	12
CLA	0	0	0	0	122	464	165	624	336	315	1302	0	0	0	2400	0	5728	33.7	15
CFI	0	0	0	0	244	174	55	156	0	270	0	1521	105	270	120	1358	4273	25.1	16
CF	0	0	70	65	0	812	440	312	144	0	294	390	3640	180	216	14	6577	38.7	14
DGW	0	0	0	65	610	464	330	5772	1248	315	0	0	0	0	0	14	8818	51.9	7

Table 8.6 Rankings of effects of climatic shocks by farmers

Factors	Rank							Total
	1	2	3	4	5	6	7	
DCY	88	37	0	0	0	0	45	170
LA	21	18	32	24	56	2	17	170
LI	61	77	7	9	16	0	0	170
FIS	0	20	18	89	11	28	4	170
DL	0	9	79	28	38	2	14	170
DC	0	9	20	16	46	69	10	170
DH	0	0	14	4	3	69	80	170
Total	170	170	170	170	170	170	170	1190
Garret-score	79	66	57	50	43	34	21	

Table 8.7 Garret ranking of effects of climatic shocks

Factors	Rank							Total score	Average score	Rank
	1	2	3	4	5	6	7			
	Total score									
DCY	6952	2442	0	0	0	0	945	10,339	60.82	2
LA	1659	1188	1824	1200	2408	68	357	8704	51.20	3
LI	4819	5082	399	450	688	0	0	11,438	67.28	1
FIS	0	1320	1026	4450	473	952	84	8305	48.85	5
DL	0	594	4503	1400	1634	68	294	8493	49.96	4
DC	0	594	1140	800	1978	2346	210	7068	41.58	6
DH	0	0	798	200	129	2346	1680	5153	30.31	7

8.4.3 Mitigation Strategies Used by Farmers

Farmers in both the strategies followed several mitigation strategies to tide over the negative impacts of climate change (Table 8.8). It shows that out of 70 farmers in Madhya Pradesh, 65 of them (93%) have raised crop only in one season, 68 farmers (97%) have sold livestock and 44 farmers (63%) have leased part of land and 50% of the farmers have shifted to other crops. Also, all the 70 farmers (100%) in Madhya Pradesh borrowed money and received assistance from government. The low income from crops have forced 53 farmers (75%) to go for less food consumption, 61 farmers (81%) have shifted to non-farm employment and 64 (91%) of farmers have out migrated to cities. In the case of Maharashtra, out of 100 farmers, 50 farmers (50%) shifted to cities, 32 farmers (32%) left land fallow, 30 farmers (30%) received support from Government, 25 farmers (25%) shifted to non-farm employment. The above discussion brings forth the severity of the effect of climate change on the livelihood of the farmers.

Table 8.8 Frequency distribution of mitigation strategies followed by farmers

Strategy	Madhya Pradesh	Maharashtra	Total
Did nothing	18	10	28
Left land fallow	52	32	84
Sold part of land	35	0	35
Leased out part of land	44	0	44
Restricted number of seasons for rice growing to 1	65	11	76
Sold livestock (cows, buffalos)	68	0	68
Shifted to other crops	35	15	50
Borrowed money from relatives/others	70	10	80
Received assistance from government/NGOs	70	30	100
Less food consumption	53	20	73
Shifted to non-farm employment	61	25	86
Out migration to cities	64	50	114

8.4.4 *Opinions of Farmers on Government Initiatives*

From time to time both Central and State governments implement several mitigation and long term measures to offset the negative effects of climate change on agriculture. With a view to know the awareness of these measures among the farmers, they were questioned on seven government initiatives (as described in the methodology chapter). They were asked to state their replies as either 'yes', or 'no' or 'not aware'. Figure 8.4 gives a pictorial representation of the awareness on these initiatives by the farmers belonging to the entire region.

In general, the government initiatives are not popular among farmers. Majority of the farmers said "No" or "not aware". However, among the farmers who are aware of the programs, about 38% of the farmers know the introduction of new crop varieties, 27% have good understanding of water regulation by government through irrigation department and only 11% of the farmers have understanding on the subsidised inputs (like fertilizers, drip irrigation system and machinery) and establishment of village knowledge centres. Only 9% of the farmers participated in the field demos of the listed adaptation practices. Further, weather based cropping insurance schemes are not popular among the farmers as only 1% of them is aware of it. Thus, it could be concluded that except (i) introduction of new crop varieties, and (ii) water regulation by government through irrigation department, all other government initiatives are not much popular among farmers. This conclusion calls for formulating suitable popularisation mechanisms by the government. Table 8.9 provides a summary of the opinions on government initiatives.

Chi-square test was applied to check if the opinions differ across the states and the results show that the opinions of the farmers on government initiatives differ strongly

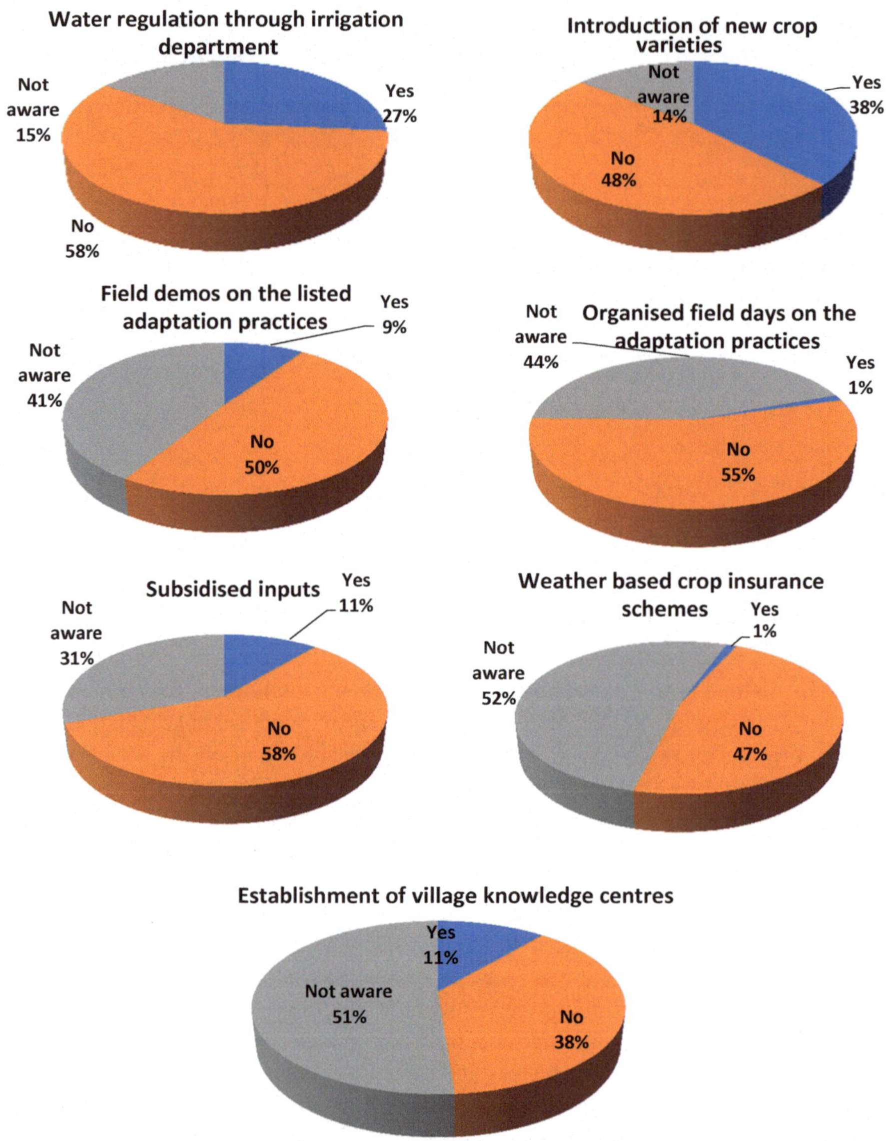

Fig. 8.4 Opinions of farmers of Central region on various government initiatives

Table 8.9 Farmers opinions on government initiatives

Govt. initiations	Madhya Pradesh			Maharashtra			Total			Chi-square	Df	*p*-value
	Yes	No	Not aware	Yes	No	Not aware	Yes	No	Not aware			
Water regulation through irrigation department	45	20	5	0	79	21	45	99	26	87.44	2	<0.00001
Introduction of new crop varieties	64	2	4	0	80	20	64	82	24	148.18	2	<0.00001
Field demos on the listed adaptation practices	0	18	52	16	66	18	16	84	70	56.41	2	<0.00001
Organised field days on the adaptation practices	0	15	55	2	78	20	2	93	75	57.51	2	<0.00001
Subsidized inputs (fertilizer, drip and machinery)	4	20	46	15	79	6	19	99	52	69.16	2	<0.00001
Weather based crop insurance schemes	2	0	68	0	80	20	2	80	88	106.19	2	<0.00001
Establishment of village knowledge centres, climate cell etc.	1	0	69	18	64	18	19	64	87	107.15	2	<0.00001

NA not aware; The *p*-values of the chi-square test show that the opinions of farmers strongly differ between states

between the two states. This may be because the government initiatives depend on local natural resources availability and other factors.

8.5 Analysis of the Effect of Climate Change on Rice Yield

8.5.1 Data

The objective of the study is to estimate the effects of climate variables on the yield of rice crop using historical data. Hence, we used a panel data set of yield and climate variables pertaining to 31 years from 1980 to 2010. The data on temperature corresponds to the growing season of rice variety in the respective states and the time series data on rainfall was collected for the entire year to account for rainfall falling directly on the crop and inter-seasonal water accumulation within a year (Iski and Devadoss 2006; Kim and Pang 2009). The summary statistics of all the variables are discussed below:

(a) **Yield (tons per ha)**

The yield data (Table 8.10; Fig. 8.5) shows an upward trend in Madhya Pradesh as the coefficient of time is positive and strongly significant. But the yield data for Maharashtra do not exhibit such trend even though it had higher yield (per ha) than that of Madhya Pradesh during most of the years (Fig. 8.5).

(b) **Maximum Temperature** (°C)

The maximum temperature had an overall average of 32.14 °C. Maharashtra had the highest mean maximum temperature of 32.85 °C. Further for both states, the data do not exhibit any trend and the coefficients of time are not significant (Fig. 8.6; Table 8.11).

(c) **Minimum Temperature** (°C)

The minimum temperature had an overall average of 19.51 °C. Madhya Pradesh had the lowest mean minimum temperature of 18.61 °C. Also, the minimum temperature of Maharashtra was higher than that of Madhya Pradesh for all the 31 years

Table 8.10 Descriptive statistics of yield and trend equations (tons/ha)

State	*N*	Mean	Min	Max	Std. Dev.	Trend equation	R^2
Madhya Pradesh	31	2.11	0.96	3.82	0.91	$Y = 0.736 + 0.086^{***}$ t	0.736
Maharashtra	31	3.02	1.24	4.34	0.774	$Y = 3.105 - 0.0051$ t	0.004
Overall	62	2.57	0.96	4.34	0.957	–	–

***Significant at 1% level; *D*1 and *D*2 are dummy variables representing Punjab and Haryana states

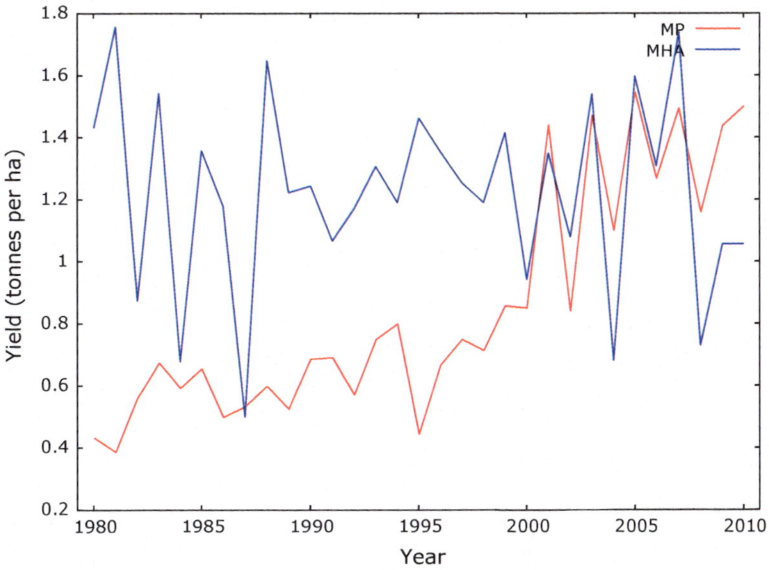

Fig. 8.5 Trend in yield of rice in Central Region States of India

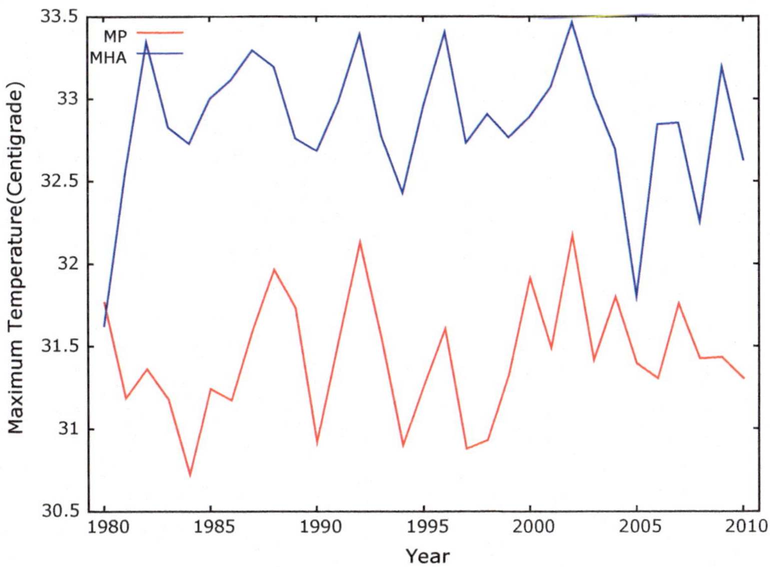

Fig. 8.6 Trend in maximum temperature in Central Region of India

Table 8.11 Descriptive statistics of maximum temperature (°C)

State	*N*	Mean	Min	Max	Std. Dev.	Trend equation	R^2
Madhya Pradesh	31	31.43	30.73	32.17	0.065	$Y = 31.33 + 0.0064T$	0.03
Maharashtra	31	32.85	31.63	33.46	0.417	$Y = 32.87 - 0.001T$	0.001
Overall	62	32.14	30.73	33.46	0.811	–	–

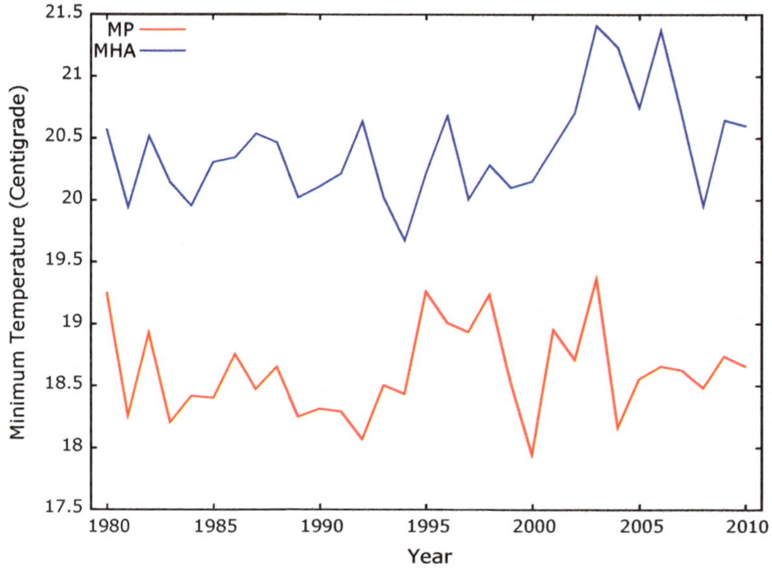

Fig. 8.7 Trend in minimum temperature in Central Region of India

Table 8.12 Descriptive statistics of minimum temperature (°C)

State	*N*	Mean	Min	Max	Std. Dev.	Trend equation	R^2
Madhya Pradesh	31	18.61	17.94	19.37	0.368	$Y = 18.536 + 0.0048T$	0.014
Maharashtra	31	20.40	19.67	21.41	0.417	$Y = 20.099 + 0.019^{**}T$	0.179
Overall	62	19.51	17.94	21.41	0.986	–	–

** denotes significance at 5% level

(Fig. 8.7). Further, the data of Maharashtra showed an increasing trend as the coefficients of time was statistically significant (Table 8.12) .

Table 8.13 Descriptive statistics of rainfall (mm)

State	N	Mean	Min	Max	Std. Dev.	Trend equation	R^2
Madhya Pradesh	31	1531	883	2636	377	Y = 1555.6 − 1.55T	0.001
Maharashtra	31	1224	771	1968	261	Y = 1215.7 + 0.496T	0.0003
Overall	62	1377	771	2636	357	–	–

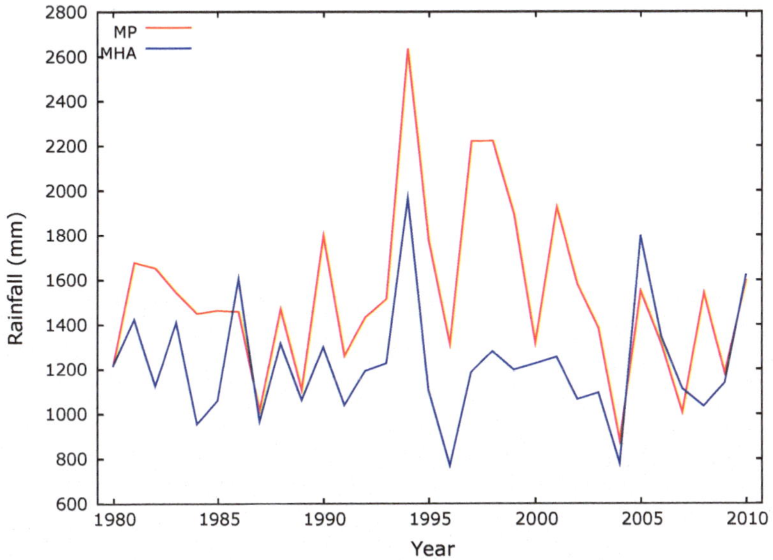

Fig. 8.8 Trend in rainfall in Central Region of India

(d) **Rainfall** (mm)

The region had an overall mean rainfall of 1377 mm with a standard deviation of 357 mm (Table 8.13). Annual rainfall of Madhya Pradesh was higher than that of Maharashtra during most of the years (Fig. 8.8). The data for both states did not exhibit any trend over time as the coefficients of trend were statistically non significant.

8.5.2 *Test of Normality of Yield Data*

The normality of yield data was tested using Shapiro-Wilk W test and Lilliefors test. The tests showed that the null hypothesis that the data is normal cannot be rejected. The computed value of Shapiro-Wilk W test was 0.9450 with a p-value of 0.0769.

Table 8.14 Panel root test results using Fisher's type Tests

Variable	Test statistic for individual effects and no trend	p-value	Decision	Test statistic for individual effect with trend	p-value	Decision
Yield	9.3555	0.0528*	Reject H_0	9.8813	0.0425**	Reject H_0
Maximum temperature	18.4207	0.001***	Reject H_0	18.4207	0.001***	Reject H_0
Minimum temperature	16.5881	0.0023***	Reject H_0	17.0344	0.0019***	Reject H_0
Rainfall	18.4207	0.001***	Reject H_0	18.4207	0.001***	Reject H_0

Since this value is greater than 0.05, the null hypothesis was not rejected. Similarly, the test statistic for Lilliefors test was 1.2956 with a p-value of 0.1198 which is also greater than 0.05. So, both tests did not reject the null hypothesis that yield data is normal.

8.5.3 Panel Root Test: Fisher's Type Tests

Since data on rice yield for the 2 States for 30 years is available as a panel data set, it was subjected to panel root testing before modelling. As stated in the methodology Chap. 4, Fisher's type panel root test was applied and the results are presented in Table 8.14.

The test results confirmed that all the variables in panel data do not have unit roots and hence they are stationary.

8.5.4 Estimation of Just-Pope Production Function

As discussed in the methodology chapter, to account for both mean and variance in yield, Just-Pope Production function (Just and Pope 1978 & 79) with yield as dependent variable and maximum and minimum temperatures, rainfall, their squares and interaction terms and dummy variables for states were used as independent variables. Table 8.15 gives the results of the estimated empirical models.

The linear model was selected based on AIC criterion. It has the least AIC value. For the mean function maximum temperature has a strong negative effect on rice yield. It shows that every 1 °C increase in maximum temperature will reduce the

Table 8.15 Estimation results of Just-Pope production function

Mean function	Linear		Quadratic	
	Coefficient	S. error	Coefficient	S. error
Constant (β_0)	5.9449	3.0164	2.3181	106.2396
Trend (β_1)	0.0299***	0.0035	0.0301	0.0036
Max$T$ (β_2)	−0.2065***	0.0747	2.9486	7.6797
Min$T$ (β_3)	0.0879	0.0816	−4.1877	4.8272
Rain$T$ (β_4)	−0.000079	0.000086	−0.0094	0.0098
Max$T$2 (β_5)	–	–	−0.1190	0.1406
Min$T$2 (β_6)	–	–	−0.0730	0.0966
Rain$T$2 (β_7)	–	–	0.0000	0.0000
Max$T$Min$T$ (β_8)	–	–	0.2170	0.1409
Max$T$Rain$T$ (β_9)	–	–	0.0003	0.0004
Min$T$RainT (β_{10})	–	–	0.0001	0.0003
MP (β_{11})	−0.5866***	0.2081	−0.6413	0.2456
Variance				
Constant	−8.9320	15.9732	−17.1926	15.1368
Time	0.0207	0.0241	0.0340	0.0301
Max$T$	−1.0253	0.7755	−0.5949	0.6432
Min$T$	1.9960***	0.6367	1.6823	0.4858
Rain$T$	−0.000051	0.0008478	0.0001	0.0008
LL	−4.9068		−1.8315	
AIC	**31.8136**		37.6630	
BIC	55.2120		73.8243	

***, **, *Significant at 1, 5 and 10% levels

average yield per ha by 0.21 tonnes. Trend has a positive coefficient and it significant at 1% level. Maximum temperature and rainfall are statistically non-significant. The coefficient for Maharashtra (dummy variable) is negative and significant implying that, on an average, difference in mean yield (per ha) between these two states is 0.5866 assuming all other variables fixed. So, Maharashtra has lower productivity compared to that of Madhya Pradesh. The minimum temperature had strong positive effect on yield variability. Its coefficient in the variance function is 1.996 which shows that yield variability increases by 1.996 for every 1 °C increase in minimum temperature. Thus, minimum temperature is risk increase variable for rice yield.

The elasticities for all climatic variables at mean levels are presented in Table 8.16 for both mean function and variance function. The elasticity for the maximum temperature is −6.38 implying that 1% increase in maximum temperature will induce a decrease of 6.38% in mean yield. Similarly, the elasticity coefficient for minimum temperature is 1.65 and so 1% increase in minimum temperature will increase the

Table 8.16 Elasticities of climatic variables at means

Yield function	Climate variable	Elasticity
Mean	Maximum temperature	−6.37967
	Minimum temperature	1.64565
	Rainfall	−0.104689
Variance	Maximum temperature	−32.7449
	Minimum temperature	38.8409
	Rainfall	−0.0688501

Table 8.17 Climate change impacts on percentage of mean yield and yield variability

Scenario	Period	Percentage change in				
		*T*max	*T*min	Rainfall	Mean yield	Yield variability
RCP4.5	2021–2050	4.1615	8.2665	18.2199	−14.85	183.55
	2071–2100	8.6017	16.1215	5.5231	−28.92	344.13
RCP8.5	2021–2050	4.5133	10.0345	34.9925	−15.94	239.55
	2071–2100	14.29309	29.36223	74.89477	−50.71	667.27

mean yield by 1.65%. The elasticity coefficient for rainfall is −0.10 which is very small and so 1% increase in rainfall will induce a small decrease of 0.10% in mean yield. When all the three variables are simultaneously increased by 1%, we can expect a substantial decrease of 4.8% in yield.

For the variance function, the elasticity coefficient of minimum temperature is 38.84 implying an increase in variability in yield of 38.8 for 1% increase in minimum temperature. Similar interpretations can be given for the elasticities of maximum temperature and rainfall, even though their coefficients are not significant. All together, we can expect an increase of 6.0% in yield when all the three variables are increased by 1%. So, we conclude that all the three variables, put together are risk increasing in this sense.

8.5.5 Climate Change Scenarios and Rice Yield Changes

The estimated elasticities given in Table 8.16 can further be used to estimate the impacts of climate change on mean yield and yield variability. For this purpose, two climate change scenarios, viz., RCP4.5 and RCP8.5 (as discussed in Chap. 3) were considered (Table 8.17).

Table 8.17 shows that, as per RCP4.5 and RCP8.5 scenarios, there will be substantial reduction in percentages of rice yield during the periods 2021–50 and 2071–2100. Under both scenarios, percentage reduction in the later period will be more. The percentage reduction in mean yield varies from 14.85 to 15.94 during the first period as per the two scenarios and as per RCP8.5 scenario it varies from 28.92 to 50.71% during the second period.

Yield variability as estimated by RCP8.5 scenario is more than that predicted by RCP4.5 scenario for both periods. The first scenario predicts an increase of 184% yield variability during 2021–2050 while the second scenario predicts an increase of 240% for the same period. For the second period, RCP4.5 predicts an increase of 344% in yield variability whereas RCP8.5 scenario predicts an increase in yield variability of 667%.

8.6 Summary

Farmers in the region perceive drought followed by increase in pest damage and untimely rain as the most important climate change events. Farmers ranked loss of income as the most important climate change shock followed by decline in crop yield and loss of assets. Farmers in both the states followed different mitigation strategies. Farmers, in general, are not aware of government initiatives like field demos and weather based cropping insurance schemes as only less than 40% of them know about them. Hence it is very important that state governments need to take suitable actions to popularise these initiatives among farmers.

Historical data on rice yield shows an upward trend in Madhya Pradesh, but the data for Maharashtra had no such trend over years. Maharashtra had the higher maximum and minimum temperatures than that of Madhya Pradesh. Annual rainfall of Madhya Pradesh was higher than that of Maharashtra during most of the years. Maximum temperature had a strong negative effect on mean rice yield while minimum temperature had strong positive effect on yield variability. So the minimum temperature is risk increasing variable for rice yield. There is no nonlinear effect of climate variables on rice yield which is in contrast to the results obtained in other regions.

The two climate change scenarios viz., RCP4.5 and RCP8.5, predict substantial reduction in percentages of rice yield during the periods 2021–2050 and 2071–2100. Yield variability as estimated by RCP8.5 scenario is more than that predicted by RCP4.5 scenario for both periods.

Appendix

Table 8.18 Factors influencing farmers perception of climate change—Garrett ranking: Madhya Pradesh

Factors	Rank																Total
	1	2	3	4	5	6	7	8	9	10	11	12	13	14	15	16	
Drought	21	12	7	11	12	4	1	1	1	0	0	0	0	0	0	0	70
Hailstorm	0	0	0	1	0	0	0	0	0	0	0	13	20	8	15	13	70
Flood	0	0	0	0	0	1	13	14	9	0	5	0	21	6	0	1	70
AD	0	0	0	0	0	2	13	12	9	0	0	0	7	21	0	6	70
IPD	0	16	0	0	11	2	6	7	18	4	0	1	4	0	1	0	70
IFM	0	0	0	0	0	0	0	0	0	5	0	13	13	12	13	14	70
UR	8	6	15	12	10	9	10	0	0	0	0	0	0	0	0	0	70
IW	5	0	13	14	1	4	4	2	4	12	10	1	0	0	0	0	70
TFH	0	0	0	0	0	18	25	10	10	4	3	0	0	0	0	0	70
TFL	0	0	5	6	7	2	8	22	8	4	8	0	0	0	0	0	70
SSDI	0	0	0	0	0	2	6	7	0	14	0	20	10	10	1	0	70
CCP	0	0	0	7	14	0	1	5	0	0	10	12	9	4	1	7	70
CLA	0	0	0	0	2	8	3	12	7	7	31	0	0	0	0	0	70
CFI	0	0	0	0	0	0	0	2	0	6	0	39	3	9	5	6	70
CF	0	0	1	1	0	14	7	6	3	0	7	10	5	6	9	1	70
DGW	0	0	0	1	10	8	6	12	26	7	0	0	0	0	0	0	70
Total	34	34	41	53	67	74	103	112	95	63	74	109	92	76	45	48	1120
Garret-score	86	76	70	65	61	58	55	52	48	45	42	39	35	30	24	14	

Table 8.19 Rankings of factors representing climate change using Garrett ranking: Madhya Pradesh

Factors	Rank																Total score	Average score	Rank
	1	2	3	4	5	6	7	8	9	10	11	12	13	14	15	16			
	Total score																		
Drought	1806	912	490	715	732	232	55	52	48	0	0	0	0	0	0	0	5042	72.0	1
Hailstorm	0	0	0	65	0	0	0	0	0	0	0	507	700	240	360	182	2054	29.3	16
Flood	0	0	0	0	0	58	715	728	432	0	210	0	735	180	0	14	3072	43.9	9
AD	0	0	0	0	0	116	715	624	432	0	0	0	245	630	0	84	2846	40.7	13
IPD	0	1216	0	0	671	116	330	364	864	180	0	39	140	0	24	0	3944	56.3	4
IFM	0	0	0	0	0	0	0	0	0	225	0	507	455	360	312	196	2055	29.4	15
UR	688	456	1050	780	610	522	550	0	0	0	0	0	0	0	0	0	4656	66.5	2
IW	430	0	910	910	61	232	220	104	192	540	420	39	0	0	0	0	4058	58.0	3
TFH	0	0	0	0	0	1044	1375	520	480	180	126	0	0	0	0	0	3725	53.2	6
TFL	0	0	350	390	427	116	440	1144	384	180	336	0	0	0	0	0	3767	53.8	5
SSDI	0	0	0	0	0	116	330	364	0	630	0	780	350	300	24	0	2894	41.3	12
CCP	0	0	0	455	854	0	55	260	0	0	420	468	315	120	24	98	3069	43.8	10
CLA	0	0	0	0	122	464	165	624	336	315	1302	0	0	0	0	0	3328	47.5	8
CFI	0	0	0	0	0	0	0	104	0	270	0	1521	105	270	120	84	2474	35.3	14
CF	0	0	70	65	0	812	385	312	144	0	294	390	175	180	216	14	3057	43.7	11
DGW	0	0	0	65	610	464	330	624	1248	315	0	0	0	0	0	0	3656	52.2	7

Table 8.20 Factors influencing farmers perception of climate change—Garrett ranking: Maharashtra

Factors	Rank																Total
	1	2	3	4	5	6	7	8	9	10	11	12	13	14	15	16	
Drought	24	23	13	1	0	0	0	0	0	38	1	0	0	0	0	0	100
Hailstorm	17	2	4	1	20	0	0	0	1	53	0	0	0	2	0	0	100
Flood	29	10	3	3	0	0	0	0	0	2	0	0	0	53	0	0	100
AD	0	9	25	14	2	0	0	0	11	6	0	11	0	21	0	1	100
IPD	20	49	20	3	3	0	0	0	2	0	0	1	0	0	0	2	100
IFM	9	5	26	23	10	1	0	0	0	1	1	0	0	23	0	1	100
UR	1	1	9	38	47	1	0	0	0	0	0	3	0	0	0	0	100
IW	0	1	0	5	8	0	0	0	86	0	0	0	0	0	0	0	100
TFH	0	0	0	0	1	87	1	0	0	0	1	7	0	0	0	3	100
TFL	0	0	0	0	0	0	97	0	0	0	0	0	1	1	0	1	100
SSDI	0	0	0	1	1	1	0	0	0	0	97	0	0	0	0	0	100
CCP	0	0	0	11	4	7	0	0	0	0	0	78	0	0	0	0	100
CLA	0	0	0	0	0	0	0	0	0	0	0	0	0	0	100	0	100
CFI	0	0	0	0	4	3	1	1	0	0	0	0	0	0	0	91	100
CF	0	0	0	0	0	0	1	0	0	0	0	0	99	0	0	0	100
DGW	0	0	0	0	0	0	0	99	0	0	0	0	0	0	0	1	100
Total	100	100	100	100	100	100	100	100	100	100	100	100	100	100	100	100	1600
Garret-score	86	76	70	65	61	58	55	52	48	45	42	39	35	30	24	14	

Table 8.21 Rankings of factors representing climate change using Garrett ranking: Maharashtra

Factors	Rank																	Average score	Rank
	1	2	3	4	5	6	7	8	9	10	11	12	13	14	15	16	Total score		
	Total Score																		
Drought	2064	1748	910	65	0	0	0	0	0	1710	42	0	0	0	0	0	6539	65.4	2
Hailstorm	1462	152	280	65	1220	0	0	0	48	2385	0	0	0	60	0	0	5672	56.7	5
Flood	2494	760	210	195	0	0	0	0	0	90	0	0	0	1590	0	0	5339	53.4	8
AD	0	684	1750	910	122	0	0	0	528	270	0	429	0	630	0	14	5337	53.4	9
IPD	1720	3724	1400	195	183	0	0	0	96	0	0	39	0	0	0	28	7385	73.9	1
IFM	774	380	1820	1495	610	58	0	0	0	45	42	0	0	690	0	14	5928	59.3	4
UR	86	76	630	2470	2867	58	0	0	0	0	0	117	0	0	0	0	6304	63.0	3
IW	0	76	0	325	488	0	0	0	4128	0	0	0	0	0	0	0	5017	50.2	11
TFH	0	0	0	0	61	5046	55	0	0	0	42	273	0	0	0	42	5519	55.2	6
TFL	0	0	0	0	0	0	5335	0	0	0	0	0	35	30	0	14	5414	54.1	7
SSDI	0	0	0	65	61	58	0	0	0	0	4074	0	0	0	0	0	4258	42.6	13
CCP	0	0	0	715	244	406	0	0	0	0	0	3042	0	0	0	0	4407	44.1	12
CLA	0	0	0	0	0	0	0	0	0	0	0	0	0	0	2400	0	2400	24.0	15
CFI	0	0	0	0	244	174	55	52	0	0	0	0	0	0	0	1274	1799	18.0	16
CF	0	0	0	0	0	0	55	0	0	0	0	0	3465	0	0	0	3520	35.2	14
DGW	0	0	0	0	0	0	0	5148	0	0	0	0	0	0	0	14	5162	51.6	10

Chapter 9
Climate Change Impacts on Rice Yield in Southern Region of India

K. Palanisami, Krishna Reddy Kakumanu, Udaya Sekhar Nagothu and C. R. Ranganathan

9.1 Introduction

The southern region for the present study covered three states: Tamil Nadu (TN), Andhra Pradesh (AP) and Karnataka (KA), and in each state, one district where rice is predominantly grown was selected, viz. Thanjavur, West Godavari and Shimoga, respectively. As stated in the methodology chapter, the present study is based on (i) primary data: farm survey with 313 farmers constituting 112 farmers from Tamil Nadu, 101 from Andhra Pradesh and 100 from Karnataka and (ii) secondary data: time series data (1981–2010) on productivity (yield) of rice and climate variables from the three districts. The location map of the study region is given in Fig. 9.1.

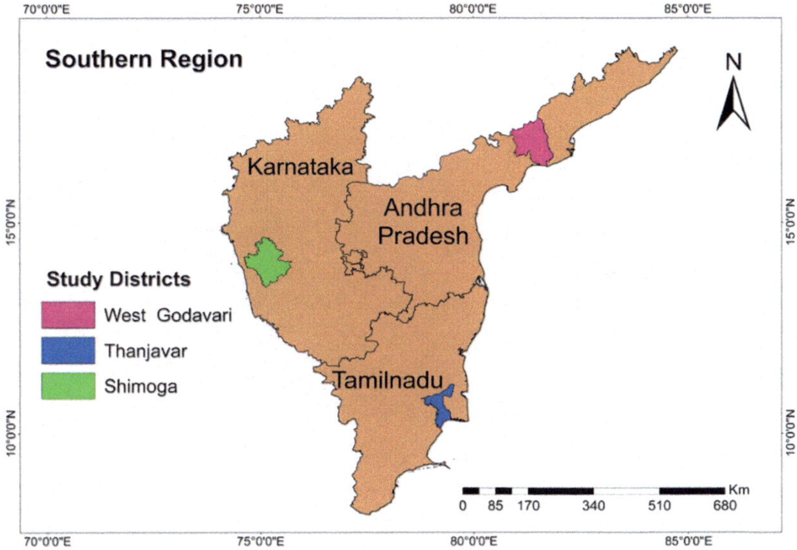

Fig. 9.1 Study regions

K. Palanisami et al., *Climate Change and Future Rice Production in India*, India Studies in Business and Economics,
https://doi.org/10.1007/978-981-13-8363-2_9

9.2 Socio-economic Profiles of the Farmers

Out of the 313 farmers surveyed, 302 farmers were males. The typology of crop irrigation differs among the states (Table 9.1). In Tamil Nadu, crop with supplemental irrigation is followed by all the farmers while 100% of the farmers surveyed in Andhra Pradesh and 78% in Karnataka have irrigated crop-based typology.

Overall, irrigated crop-based typology is followed by 57% of the farmers and 37% follow crop with supplemental irrigation and the rest 6% follow a combination of typologies as shown in Fig. 9.2 and Table 9.1.

The average age of the farmers in the three states was nearly equal and is respectively 48.5, 47.7 and 51.2 years. About 27% of the farmers had no formal education, 29% have studied up to fifth standard, 32% up to primary level and only 12% have the education of secondary and above (Fig. 9.3).

Among the states, Tamil Nadu had the highest percentage (30) of farmers who had done education up to the level of secondary and above while Andhra Pradesh had the highest percentage (55) of farmers with no formal education. Karnataka had the highest percentage (60) of farmers who had done up to primary level.

Table 9.1 Cropping typologies of farmers in the three states of southern region

State/typology	Irrigated crop-based (1)	Crop with Supplemental Irrigation (2)	Dry land-based (3)	(1) and (2)	(1) and (3)	(1), (2) and (3)	Total
Tamil Nadu	0	112	0	0	0	0	112
Andhra Pradesh	101	0	0	0	0	0	101
Karnataka	78	3	0	13	4	2	100
Total	179	115	0	13	4	2	313

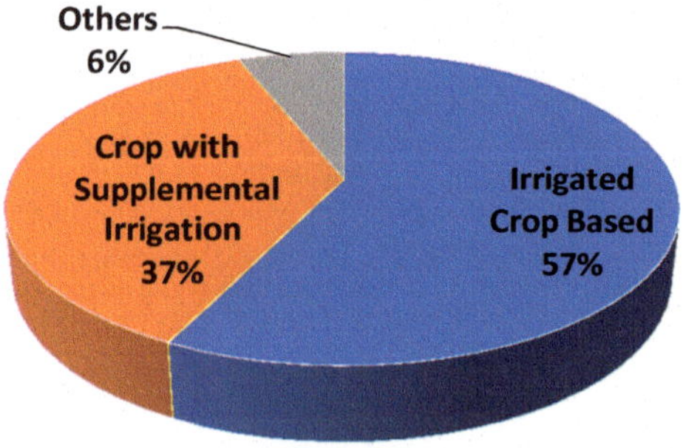

Fig. 9.2 Typologies of households in southern region

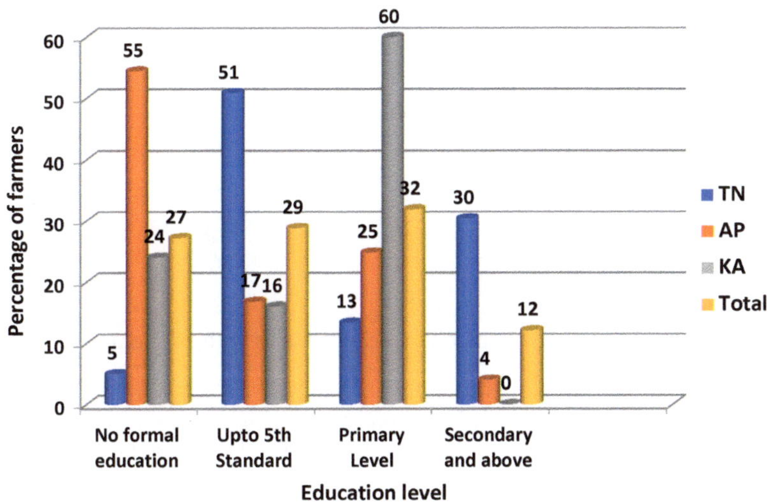

Fig. 9.3 Educational status of surveyed farmers

9.3 Farm Information

Farmers in the study area have wells as the main source of irrigation to their lands. Irrigation is provided to the crops by adopting different irrigation methods like drip, sprinkler and improved gravity irrigation. Mostly, rice is grown under assured water supply situations like canals and tanks where one rice crop is grown during Kharif season using gravity irrigation methods. In the rabi season and in those areas with well irrigation sources, farmers grow irrigated dry crops using both gravity and micro-irrigation methods.

(a) **Physical assets:**

(i) Land owned: Farmers in Tamil Nadu and Karnataka own wet, irrigated and dry lands, whereas farmers in Andhra Pradesh own only irrigated land. Table 9.2 gives the summary statistics of the land ownership pattern:

Table 9.2 Land owned by farmers in the three states

State	Average land area owned (ha)		
	Wet	Irrigated	Dry
Tamil Nadu	0.58	0.83	0.47
Andhra Pradesh	0.00	1.95	0.00
Karnataka	1.50	1.47	0.02
Total	0.68	1.40	0.17

Table 9.3 Distribution of number of borewells owned across states. (no. of farmers)

State	No. of borewells owned				Total
	1	2	3	4	
Tamil Nadu	32	68	11	1	112
Andhra Pradesh	101	0	0	0	101
Karnataka	99	1	0	0	100
Total	232	69	11	1	313

Overall, the region is predominated by irrigated land with an average area of 1.4 ha per farmer.

(ii) **Wells owned**: No farmer possesses open well and they own only borewells due to the declining groundwater table in the region. The following table gives the frequency distribution of a number of borewells owned by the farmers (Table 9.3).

The table shows that most of the farmers (74%) own at least one borewell and 22% of farmers own 2 borewells.

(iii) **Sources of irrigation**: Farmers have three sources of irrigation, viz. canal, tanks and well. In the case of the sample farmers studied, all the sample farmers in Tamil Nadu use both canal and well for irrigation, whereas all the farmers in Andhra Pradesh and Karnataka depend on the canal for irrigation mostly for growing rice in the kharif season.

(iv) **Investment in irrigation**: Most of the irrigation investments in the regions are for digging and drilling new wells. Farmers in several locations also use drip and sprinkler irrigation for irrigating non-rice crops. Several farmers also made investments in farm ponds to recharge groundwater and also for supplemental irrigation of the irrigated dry crops.

(v) **Area under rice**: Farmers in all the three states have cultivated rice only during the kharif season and some in rabi season. During kharif season, the average area under rice in the three states were respectively 0.58, 1.95 and 0.72 ha. Similarly, in rabi season, the average rice area in the three states were 0.83, 1.95 and 0.64 ha. Other crops are grown during the summer season by Tamil Nadu and Andhra Pradesh farmers with an average area of 0.27 and 0.31 ha while Karnataka farmers grew other crops only during kharif season and the average area is 0.73 ha.

9.4 Climate Change Events, Shocks and Their Effects as Perceived by Farmers

In order to have an in-depth understanding of farmers' perception on the climate change, its effects on agriculture and various adaptation strategies/techniques followed by the farmers, a detailed survey was conducted and the results are discussed below.

9.4.1 Perception of Farmers on Climate Change

Farmers were asked to rank the 16 events as listed in the methodology section on their perceptions of climate change. To find what the farmers perceive as the most significant effects of climate change, Garrett's ranking technique as described in Chapter 4 was used and the results are given in Tables 9.4 and 9.5. Table 9.5 shows that incidence of pests and diseases is the most important climate change event (Rank 1) perceived by rice farmers in southern region. It earned an average Garrett Score of 69.2 (Table 9.5). As given in Table 9.4, it was given Rank 1 by 93 farmers. The second and third important climate change events are respectively untimely rain and irregular weather. A similar analysis was done for individual states in the region and are given in Appendix (Tables 9.18, 9.19, 9.20, 9.21, 9.22, 9.23).

Friedman Rank Test was applied to test for differences between rankings of the climate change events (listed in Table 9.4) by the farmers. The methodology as described in Chapter 4 was followed. The calculated value of Friedman's F statistics was 1022.4 while the table value (corresponding to 15 degrees of freedom at 5% level of significance) is 24.996. This shows that the farmers did not consistently rank the events; thus, the farmers' perceptions of climate change effects differ across the sample chosen. The same conclusion was arrived at by applied Kendall's coefficient of concordance also. Similar analysis was done for individual states viz. Tamil Nadu, Andhra Pradesh and Karnataka. The study results show that there is no consistency in the rankings of the events by the farmers. Table 9.6 gives the rankings of the climate change events state-wise and also region-wise.

The above table shows that the rankings of the climate change events are not uniform across the states. So to check for agreement in rankings, the Kendall's coefficient of concordance was computed and tested for its significance. The computed value of Kendall's Coefficient, W was 0.4451 (with number of objects $n = 16$ and number of judges $m = 3$) with a χ^2 value of 20.0294 (p-value $= 0.1708$). Since the p-value is more than 0.10, we do not reject the null hypothesis and conclude that the states are concordant in ranking the climate change events.

Table 9.4 Factors influencing farmers perception of climate change—Garrett ranking: southern region

Climate change event	Rank																Total
	1	2	3	4	5	6	7	8	9	10	11	12	13	14	15	16	
Drought	38	8	5	15	23	32	17	42	2	2	0	20	32	31	22	24	313
Hailstorm	4	6	8	37	24	23	20	1	0	0	0	19	44	44	64	19	313
Flood	23	8	56	1	0	3	15	18	19	14	2	1	6	15	95	37	313
AD	17	43	18	6	10	13	18	9	50	36	20	19	7	13	19	15	313
IPD	93	46	25	52	4	10	35	29	8	1	2	2	5	0	1	0	313
IFM	3	4	2	3	31	8	9	6	13	20	43	89	55	17	3	7	313
UR	35	26	45	45	84	25	10	9	13	5	6	4	3	3	0	0	313
IW	9	37	25	54	26	66	22	26	8	6	15	1	12	6	0	0	313
TFH	35	41	24	13	22	19	16	22	13	27	12	13	34	20	2	0	313
TFL	10	0	1	15	4	13	13	11	24	25	33	21	12	40	34	57	313
SSDI	11	11	13	29	19	28	9	5	19	14	45	16	19	33	22	20	313
CCP	0	0	0	4	13	19	55	18	33	40	19	24	22	20	12	34	313
CLA	0	22	11	7	14	11	19	37	12	39	51	16	23	7	10	34	313
CFI	7	28	27	6	21	7	19	41	12	8	20	27	26	35	26	3	313
CF	8	33	24	25	17	35	16	27	56	16	36	7	6	5	1	1	313
DGW	20	0	29	1	1	1	20	12	31	60	9	34	7	24	2	62	313
Total	313	313	313	313	313	313	313	313	313	313	313	313	313	313	313	313	5008
Garrett score	86	76	70	65	61	58	55	52	48	45	42	39	35	30	24	14	

Table 9.5 Rankings of factors representing climate change using Garrett ranking: southern region

Climate change Event	Rank																Total score	Average score	Rank
	1	2	3	4	5	6	7	8	9	10	11	12	13	14	15	16			
	Total score																		
Drought	3268	608	350	975	1403	1856	935	2184	96	90	0	780	1120	930	528	336	15,459	49.4	7
Hailstorm	344	456	560	2405	1464	1334	1100	52	0	0	0	741	1540	1320	1536	266	13,118	41.9	15
Flood	1978	608	3920	65	0	174	825	936	912	630	84	39	210	450	2280	518	13,629	43.5	11
AD	1462	3268	1260	390	610	754	990	468	2400	1620	840	741	245	390	456	210	16,104	51.5	6
IPD	7998	3496	1750	3380	244	580	1925	1508	384	45	84	78	175	0	24	0	21,671	69.2	1
IFM	258	304	140	195	1891	464	495	312	624	900	1806	3471	1925	510	72	98	13,465	43.0	12
UR	3010	1976	3150	2925	5124	1450	550	468	624	225	252	156	105	90	0	0	20,105	64.2	2
IW	774	2812	1750	3510	1586	3828	1210	1352	384	270	630	39	420	180	0	0	18,745	59.9	3
TFH	3010	3116	1680	845	1342	1102	880	1144	624	1215	504	507	1190	600	48	0	17,807	56.9	4
TFL	860	0	70	975	244	754	715	572	1152	1125	1386	819	420	1200	816	798	11,906	38.0	16
SSDI	946	836	910	1885	1159	1624	495	260	912	630	1890	624	665	990	528	280	14,634	46.8	9
CCP	0	0	0	260	793	1102	3025	936	1584	1800	798	936	770	600	288	476	13,368	42.7	14
CLA	0	1672	770	455	854	638	1045	1924	576	1755	2142	624	805	210	240	476	14,186	45.3	10
CFI	602	2128	1890	390	1281	406	1045	2132	576	360	840	1053	910	1050	624	42	15,329	49.0	8
CF	688	2508	1680	1625	1037	2030	880	1404	2688	720	1512	273	210	150	24	14	17,443	55.7	5
DGW	1720	0	2030	65	61	58	1100	624	1488	2700	378	1326	245	720	48	868	13,431	42.9	13

Table 9.6 Ranking of climate change events state-wise

Climate change event	Tamil Nadu	Andhra Pradesh	Karnataka	Southern region
Drought	10	4	15	7
Hailstorm	9	10	13	15
Flood	14	2	16	11
AD	6	8	9	6
IPD	4	1	1	1
IFM	11	11	12	12
UR	2	3	5	2
IW	3	5	6	3
TFH	1	12	7	4
TFL	15	13	11	16
SSDI	12	15	3	9
CCP	16	7	8	14
CLA	7	16	10	10
CFI	13	9	2	8
CF	8	6	4	5
DGW	5	14	14	13

9.4.2 *Effects of Shocks*

Farmers were questioned on their perception about the effects of climatic shocks. They were asked to rank seven effects of climatic shocks, as described in methodology chapter. Farmers ranked declined in crop yield as the most important shock (Tables 9.7 and 9.8). Out of 313 farmers, 196 of them have ranked 1 to this effect. The second and third ranks are given to the loss of income and loss of assets, respectively. This is understandable because the decline in crop yield will reflect in income and asset generation.

9.5 Analysis of the Effect of Climate Change on Rice Yield

9.5.1 *Data*

The objective of the study is to estimate the effects of climate variables on the yield of rice crop using historical data. A panel data set of yield and climate variables pertaining to 30 years from 1981 to 2010 was used in the analysis. The data on temperature corresponds to the growing season of rice variety in the respective states and the time series data on rainfall was collected for the entire year to account for

Table 9.7 Rankings of effects of climatic shocks by farmers

Factors	Rank							Total
	1	2	3	4	5	6	7	
DCY	196	57	16	35	4	5	0	313
LA	35	18	110	38	14	33	65	313
LI	69	194	29	6	2	11	2	313
FIS	4	11	29	65	130	40	34	313
DL	7	14	28	57	90	91	26	313
DC	1	9	84	71	46	45	57	313
DH	1	10	17	41	27	88	129	313
Total	313	313	313	313	313	313	313	2191
Garrett score	79	66	57	50	43	34	21	

Table 9.8 Garrett ranking of effects of climatic shocks

Factors	Rank							Total score	Average score	Rank
	1	2	3	4	5	6	7			
	Total score									
DCY	15,484	3762	912	1750	172	170	0	22,250	71.1	1
LA	2765	1188	6270	1900	602	1122	1365	15,212	48.6	3
LI	5451	12,804	1653	300	86	374	42	20,710	66.2	2
FIS	316	726	1653	3250	5590	1360	714	13,609	43.5	5
DL	553	924	1596	2850	3870	3094	546	13,433	42.9	6
DC	79	594	4788	3550	1978	1530	1197	13,716	43.8	4
DH	79	660	969	2050	1161	2992	2709	10,620	33.9	7

rainfall falling directly on the crop and interseasonal water accumulation within a year (Iski and Devadoss 2006; Kim and Pang 2009). The summary statistics of all the variables are discussed below:

(a) **Yield (tons per ha)**

The yield data (Table 9.9; Fig. 9.4) shows an upward trend in all the three states over the years. Trend equations fitted to yield data show that the coefficient of trend for all the three states are statistically significant.

(b) **Maximum Temperature (°C)**

The maximum temperature had an overall average of 29.91 °C. Tamil Nadu had the highest mean maximum temperature of 32.97 °C followed by Andhra Pradesh (32.24 °C). Also, the variability in maximum temperature was least for Karnataka (0.17 °C) (Table 9.10). Tamil Nadu's maximum temperature is consistently higher than those of Andhra Pradesh and Karnataka. Similarly, Karnataka's temperature is consistently lower than those of the other two states with an average difference of approximately 8 °C in all the years (Fig. 9.5).

Table 9.9 Descriptive statistics of rice yield and trend equations (tons/ha)

	N	Mean	Min	Max	Std. Dev.	Trend equation	R^2
Tamil Nadu	30	2.848	1.855	3.579	0.467	Y = 2.518 + 0.0213** t	0.16
Andhra Pradesh	30	3.041	1.757	4.095	0.529	Y = 2.4162 + 0.0403*** t	0.45
Karnataka	30	2.345	1.772	3.868	0.422	Y = 1.8499 + 0.0319*** t	0.44
Overall	90	2.744	1.757	4.095	0.554		

** and *** denotes significance at 5% and 1% level

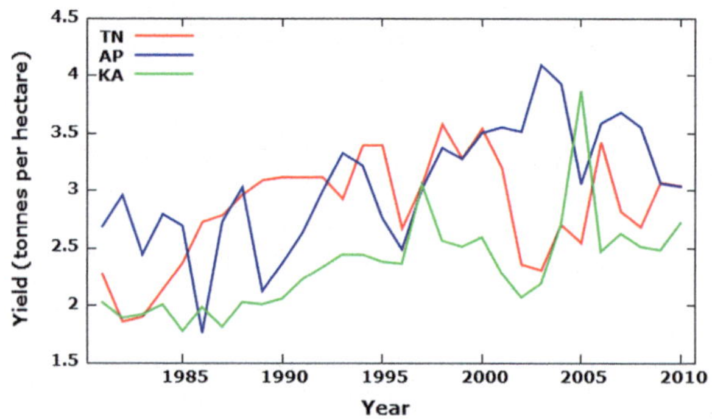

Fig. 9.4 Trend in yield of rice in southern region of India

Table 9.10 Descriptive statistics of maximum temperature (°C)

	N	Mean	Minimum	Maximum	Std. Dev.
Tamil Nadu	30	32.97	32.20	34.3	0.47
Andhra Pradesh	30	32.24	31.39	32.96	0.37
Karnataka	30	24.52	24.22	24.87	0.17
Overall	90	29.91	29.27	30.71	3.86

(c) **Minimum Temperature (°C)**

The temperature data shows that Tamil Nadu had the highest average minimum temperature (26.11 °C) compared to the other two states. Its average minimum temperature is 25.23 °C (Table 9.11). It is consistently higher than those of the other two states. Karnataka had the lowest minimum temperature with an average of 17.8 °C in all the 30 years. On the average, there was a difference of approximately 5 °C in its minimum temperature with those of Tamil Nadu and Andhra Pradesh (Fig. 9.6).

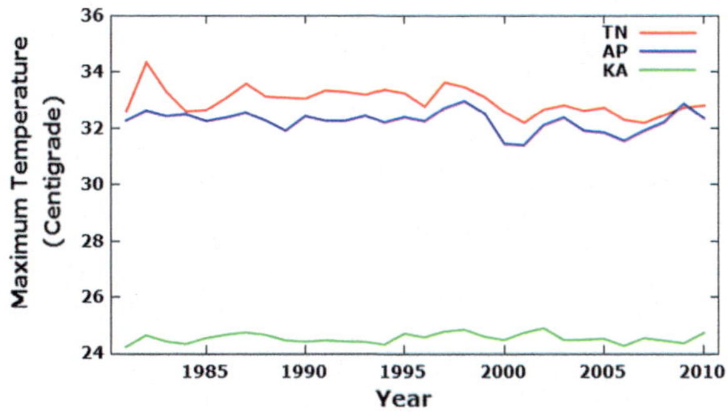

Fig. 9.5 Trend in maximum temperature in southern region of India

Table 9.11 Descriptive statistics of minimum temperature (°C)

	N	Mean	Minimum	Maximum	Std. Dev.
Tamil Nadu	30	25.23	24.59	26.11	0.43
Andhra Pradesh	30	22.95	21.94	23.72	0.37
Karnataka	30	17.79	17.48	18.13	0.17
Overall	90	21.99	17.48	26.11	3.15

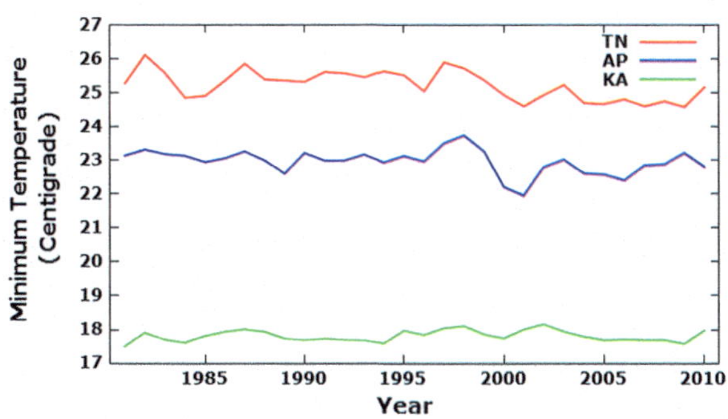

Fig. 9.6 Trend in minimum temperature in southern region of India

(d) **Rainfall (mm)**

Karnataka had the highest mean annual rainfall of 2479 mm followed by Andhra Pradesh (1149 mm). Its variability in rainfall was also highest with 593 mm. Compared with other two states, Tamil Nadu consistently received lesser rainfall in most of the years while rainfall at Karnataka was consistently higher than those of the other two states in all the years (Fig. 9.7). Thus we find that Tamil Nadu had higher temperature than Andhra Pradesh and Karnataka, whereas Karnataka had higher rainfall compared to other two states (Table 9.12).

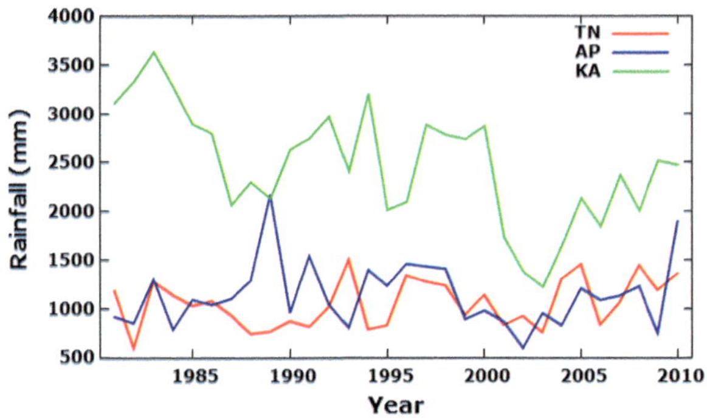

Fig. 9.7 Trend in annual rainfall in southern region of India

Table 9.12 Descriptive statistics of rainfall (mm)

	N	Mean	Minimum	Maximum	Std. Dev.
Tamil Nadu	30	1064	610	1509	246
Andhra Pradesh	30	1149	603	2185	341
Karnataka	30	2479	1226	3639	593
Overall	90	1564	603	3639	773

Table 9.13 Tests normality of yield data

Name of the test	Test statistic	*p*-value	Decision
Doornik–Hansen test	2.713	0.258	Don't reject Ho
Shapiro–Wilk W	0.980	0.190	Don't reject Ho
Lilliefors test	0.055	0.720	Don't reject Ho
Jarque–Bera test	2.334	0.311	Don't reject Ho

9.5.2 Test of Normality of Yield Data

The normality of yield data was tested using four tests. The results of the tests are given in Table 9.13. For all the tests, the null hypothesis was the data that is normally distributed. Hence based on the four tests, the null hypothesis can't be rejected.

9.5.3 Panel Root Test: Fisher's Tests

Since data on rice yield for the three states for 30 years formed a panel data set, it was subjected to panel root testing before modelling. As stated in the methodology chapter, Fisher's panel root test was applied (Table 9.14).

Table 9.14 Panel root test results using fisher's tests

Variable	Test statistic for individual effects and no trend	*p*-value	Decision	Test statistic for individual effect with trend	*p*-value	Decision
Yield	14.1114	0.0284**	Reject H_0	17.5767	0.0074***	Reject H_0
Maximum temperature	27.6310	0.0001***	Reject H_0	21.9066	0.0013***	Reject H_0
Minimum Tempera-ture	22.1173	0.0012***	Reject H_0	21.6396	0.0014***	Reject H_0
Rainfall	22.6612	0.0009***	Reject H_0	21.9646	0.0012***	Reject H_0

** and *** denotes significance at 5% and 1% level

The test results confirmed that all the variables in panel data do not have unit roots and hence they are stationary.

9.5.4 Estimation of Just–Pope Production Function

As discussed in the methodology chapter, to account for both mean and variance in yield, Just–Pope production function (Just and Pope 1978 & 79) with yield as the dependent variable and maximum and minimum temperatures, rainfall, their squares and interaction terms and dummy variables for states were used as independent variables. Table 9.15 gives the results of the estimated empirical models.

The quadratic model has the least value for AIC and so is selected. Table 9.15 shows that maximum temperature has a significant positive effect on rice yield, whereas minimum temperature has a very strong significant negative effect. Similarly, rainfall has a significant positive effect on yield. The square term corresponding to maximum temperature and two interaction terms are significant. The significance of these terms implies that nonlinear effects of climate variables are significant and so important (Figs. 9.8, 9.9 and 9.10). The coefficient of the dummy variable for Andhra Pradesh is positive and strongly significant implying that the mean yields between the states differ statistically. The positive and significant coefficient for trend implies the significance of technological progress in rice production such as improved irrigation methods, use of high yielding varieties, better control of pests and diseases.

With respect to variance function, coefficients of trend and all the three climate variables are significant. The coefficient of trend is positive implying that it is a risk-increasing variable. This means that variability in yield increases over the years. But we have already seen that the coefficient of a trend in the mean yield function is positive and significant which implies that mean yield increases over time. So, we conclude that over years increase in mean yield increases variability in yield also.

Table 9.15 Estimation results of Just–Pope production function

Mean function	Linear		Quadratic	
	Coefficient	S. error	Coefficient	S. error
Constant (β_0)	−0.8181	2.3554	46.9815	22.9983
Trend (β_1)	0.0342	0.0042	0.0388***	0.0039
Max$T$ (β_2)	−0.0352	0.2669	10.8095***	4.0932
Min$T$ (β_3)	0.2051	0.3101	−10.4427**	4.0121
Rain$T$ (β_4)	0.0000	0.0001	0.0028*	0.0016
Max$T2$ (β_5)	–	–	−0.4626**	0.2122
Min$T2$ (β_6)	–	–	−0.3063	0.2272
Rain$T2$ (β_7)	–	–	0.0000	0.0000
Max$T$Min$T$ (β_8)	–	–	0.7881**	0.4304
Max$T$Rain$T$ (β_9)	–	–	−0.0002***	0.0001
Min$T$Rain$T$ (β_{10})	–	–	0.0002	0.0001
TN (β_{11})	−0.6040	0.8050	−1.0438	1.3089
AP (β_{12})	−0.2727	0.7583	0.1012**	0.0130
Variance function				
Constant	1.2537	4.9896	−0.0831	4.1591
Time	0.0263	0.0250	0.0306*	0.0233
Max$T$	0.3751	0.2349	0.2967*	0.2310
Min$T$	−0.6369	0.2210	−0.4959**	0.2345
Rain$T$	−0.0009	0.0006	−0.0007**	0.0005
Model statistics				
LL	−17.1199		−5.6029	
AIC	58.2399		47.2058	
BIC	88.2376		92.2023	

** and *** denotes significance at 5% and 1% level

These results coincide with the findings of Palanisami et al. (2011), Ranganathan (2009), Isik and Devadoss (2006) and Anderson and Hazell (1987). Since the coefficient of maximum temperature is positive, it is a risk-increasing variable for rice yield. So, the increase in maximum temperature increases the yield variability. Coefficients of minimum temperature and rainfall are negative and statistically significant. So, these two are risk-decreasing variables.

The elasticities for all climatic variables at mean levels are presented in Table 9.16 for both mean function and variance function. The elasticity for the maximum temperature is −2.894 implying that 1% increase in maximum temperature will induce a decrease of 2.89% in mean yield. Similarly, the elasticity coefficient for minimum temperature is 3.3452 and so 1% increase in minimum temperature will increase the mean yield by 3.35%. The elasticity coefficient for rainfall is −0.1420 which is very

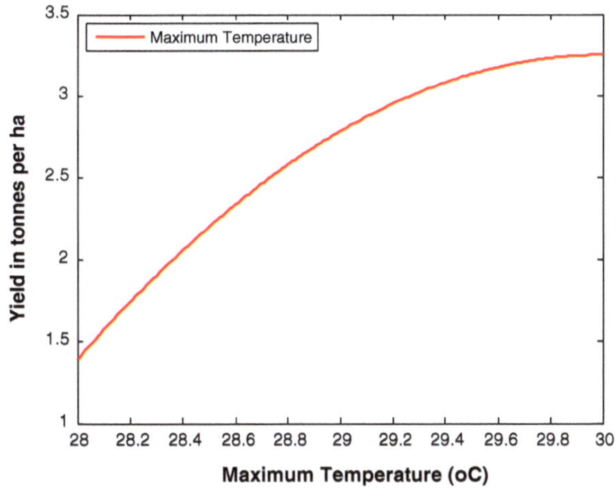

Fig. 9.8 Nonlinear relationship between yield and maximum temperature

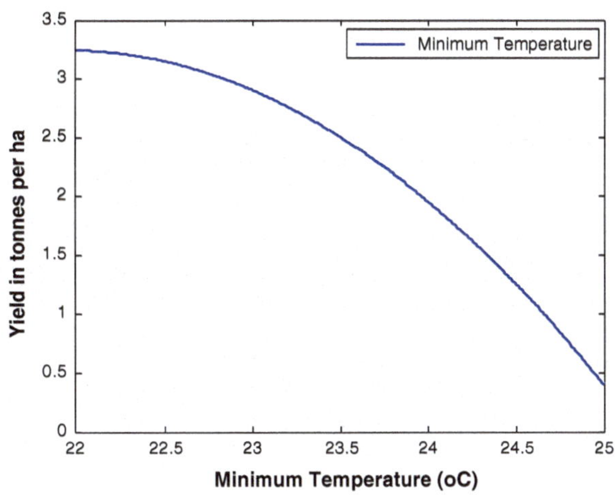

Fig. 9.9 Nonlinear relationship between yield and minimum temperature

small and so 1% increase in rainfall will induce a decrease of 0.14% in mean yield. When all the three variables are simultaneously increased by 1%, we can expect a small increase of 0.31% in yield.

For the variance function, the elasticity coefficient of maximum temperature is −5.3073 implying a reduction in variability in yield of 5.31 for 1% increase in maximum temperature. Similar meanings can be given for the elasticities of minimum temperature and rainfall. All together, we can expect a decrease of 5.19% in yield when all the three variables are increased by 1%. So, we conclude that all the three variables put together are risk-decreasing variables.

The mean function can be used to compute the optimal levels of climate variables which will give maximum yield. Using differential calculus, it can be shown that the optimal levels of Max$T$, Min$T$ and Rain$T$ are obtained by solving the following set of simultaneous equations:

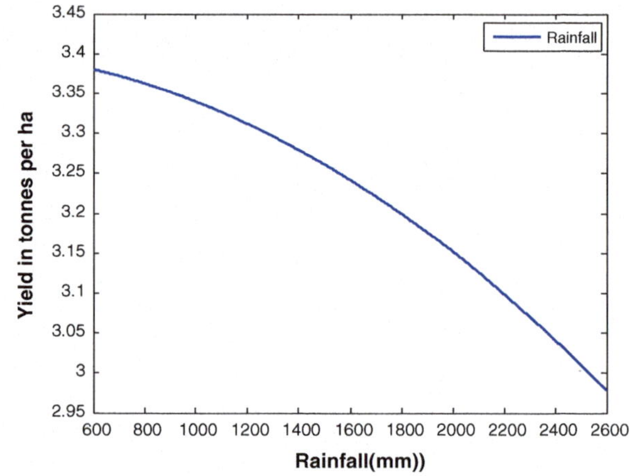

Fig. 9.10 Nonlinear relationship between yield and rainfall

Table 9.16 Elasticities of climatic variables at means

Yield Function	Climate variable	Elasticity
Mean	Maximum temperature	−2.8940
	Minimum temperature	3.3452
	Rainfall	−0.1420
Variance	Maximum temperature	−5.3073
	Minimum temperature	1.9602
	Rainfall	−1.8431

$$\frac{\partial y}{\partial(\text{Max}T)} = 2\beta_5\text{Max}T + \beta_8\text{Min}T + \beta_9\text{Rain}T = -\beta_2$$
$$\frac{\partial y}{\partial(\text{Min}T)} = \beta_8\text{Max}T + 2 * \beta_6\text{Min}T + \beta_{10}\text{Rain}T = -\beta_3$$
$$\frac{\partial y}{\partial(\text{Rain}T)} = \beta_9\text{Max}T + \beta_{10}\text{Rain}T + 2\beta_7\text{Rain}T = -\beta_4$$

Solving these equations, the optimal levels are obtained as Max$T$ = 29.96 °C; Min$T$ = 20.86 °C and Rain$T$ = 1028 mm.

9.5.5 *Climate Change Scenarios and Rice Yield Changes*

The estimated elasticities given in Table 9.17 can be used to estimate the impacts of climate change on mean yield and yield variability of yield. For this purpose, two

Table 9.17 Climate change impacts on percentage of mean yield and yield variability

Scenario	Period	Percentage change in				
		*T*max	*T*min	Rainfall	mean yield	Yield variability
RCP45	2021–2050	3.66	6.57	1.42	**11.2**	**−9.13**
	2071–2100	7.44	12.66	1.86	**20.57**	**−18.08**
RCP85	2021–2050	4.33	7.69	6.71	**12.23**	**−20.28**
	2071–2100	12.92	23.25	42.33	**34.40**	**−101.00**

climate change scenarios, viz. RCP45 and RCP85 (as discussed in Chap. 3) were considered. Table 9.17 presents the results.

The table shows that, as per RCP45 and RCP85 scenarios, the climate change will have a positive effect on rice yield in the region. There will be an increase in percentages of rice yield during the periods 2021–50 and 2071–2100 and the percentage increase in the later period (2071–2100) will be more. The percentage increase in mean yield varies from 11.2 to 12.23 during the first period and it varies from 20.57 to 34.40% during the second period.

Yield variability is expected to reduce in both periods under the two scenarios. The predicted reduction in variability is more with respect to the RCP85 scenario. The first scenario predicts a decrease in variability of 9.1% during 2021–2050 while the second scenario predicts a decrease of 20.3% for the same period. For the second period, RCP45 predicts a reduction of 18% in yield variability, whereas RCP85 scenario predicts 101% reduction in variability. Thus, we find that climate change will induce an increase in yield accompanied by a substantial reduction in yield variability in southern region.

9.6 Summary

Incidence of pests and diseases is the most important climate change event (Rank 1) perceived by rice farmers in southern region, followed by untimely rain and irregular weather. Farmers ranked decline in crop yield as the most important shock, followed by loss of income and loss of assets.

Historical rice yield data shows an upward trend in all the three states over years. Tamil Nadu registered greater maximum and minimum temperature compared with other two states throughout the study period. But Karnataka had the highest mean annual rainfall followed by Andhra Pradesh and Tamil Nadu. The analysis of relationship between yield and climate variables shows that maximum temperature had a significant positive effect on rice yield, whereas minimum temperature has very strong significant negative effect. Similarly, rainfall has a significant positive effect on yield. Also, there is strong evidence of nonlinear effects of climate variables on yield.

The two climate change scenarios, viz. RCP45 and RCP85, predict an increase in percentages of rice yield during the periods 2021–50 and 2071–2100 while yield variability is expected to reduce in both periods under the two scenarios. The percentage increase in mean yield varies from 11.2 to 12.23 during the first period and from 20.57 to 34.40 during the second period as per the two scenarios.

Appendix

Table 9.18 Factors influencing farmers' perception of climate change—Garrett ranking: Tamil Nadu

Factors	Rank																Total
	1	2	3	4	5	6	7	8	9	10	11	12	13	14	15	16	
Drought	2	6	0	0	2	24	15	24	1	1	0	11	7	19	0	0	112
Hailstorm	0	0	0	11	18	18	20	0	0	0	0	13	32	0	0	0	112
Flood	0	0	0	0	0	0	0	18	16	11	0	0	0	1	53	13	112
AD	17	2	1	6	2	5	12	5	22	25	10	3	0	2	0	0	112
IPD	14	3	5	12	2	8	34	18	7	1	2	2	4	0	0	0	112
IFM	0	3	2	0	14	7	7	3	3	3	20	33	10	7	0	0	112
UR	14	10	25	8	37	16	2	0	0	0	0	0	0	0	0	0	112
IW	5	25	14	38	6	0	0	0	1	0	11	0	12	0	0	0	112
TFH	34	38	18	8	7	5	1	0	0	0	1	0	0	0	0	0	112
TFL	0	0	0	12	0	11	3	4	6	7	0	2	0	0	13	54	112
SSDI	1	2	0	3	2	14	4	4	12	6	11	7	1	22	13	10	112
CCP	0	0	0	1	5	0	0	1	0	9	4	17	14	16	11	34	112
CLA	0	22	10	5	10	0	4	9	4	16	7	11	10	4	0	0	112
CFI	0	0	0	2	0	2	1	3	4	2	12	11	20	32	22	1	112
CF	7	1	8	6	6	2	6	16	16	11	28	0	2	3	0	0	112
DGW	18	0	29	0	1	0	3	7	20	20	6	2	0	6	0	0	112
Total	112	112	112	112	112	112	112	112	112	112	112	112	112	112	112	112	1792
Garrett score	86	76	70	65	61	58	55	52	48	45	42	39	35	30	24	14	

Table 9.19 Rankings of factors representing climate change using Garrett ranking: Tamil Nadu

Factors	Rank																Total score	Average score	Rank
	1	2	3	4	5	6	7	8	9	10	11	12	13	14	15	16			
	Total score																		
Drought	172	456	0	0	122	1392	825	1248	48	45	0	429	245	570	0	0	5552	49.6	10
Hailstorm	0	0	0	715	1098	1044	1100	0	0	0	0	507	1120	0	0	0	5584	49.9	9
Flood	0	0	0	0	0	0	0	936	768	495	0	0	0	30	1272	182	3683	32.9	14
AD	1462	152	70	390	122	290	660	260	1056	1125	420	117	0	60	0	0	6184	55.2	6
IPD	1204	228	350	780	122	464	1870	936	336	45	84	78	140	0	0	0	6637	59.3	4
IFM	0	228	140	0	854	406	385	156	144	135	840	1287	350	210	0	0	5135	45.8	11
UR	1204	760	1750	520	2257	928	110	0	0	0	0	0	0	0	0	0	7529	67.2	2
IW	430	1900	980	2470	366	0	0	0	48	0	462	0	420	0	0	0	7076	63.2	3
TFH	2924	2888	1260	520	427	290	55	0	0	0	42	0	0	0	0	0	8406	75.1	1
TFL	0	0	0	780	0	638	165	208	288	315	0	78	0	0	312	756	3540	31.6	15
SSDI	86	152	0	195	122	812	220	208	576	270	462	273	35	660	312	140	4523	40.4	12
CCP	0	0	0	65	305	0	0	52	0	405	168	663	490	480	264	476	3368	30.1	16
CLA	0	1672	700	325	610	0	220	468	192	720	294	429	350	120	0	0	6100	54.5	7
CFI	0	0	0	130	0	116	55	156	192	90	504	429	700	960	528	14	3874	34.6	13
CF	602	76	560	390	366	116	330	832	768	495	1176	0	70	90	0	0	5871	52.4	8
DGW	1548	0	2030	0	61	0	165	364	960	900	252	78	0	180	0	0	6538	58.4	5

Table 9.20 Factors influencing farmers perception of climate change—Garrett ranking: Andhra Pradesh

Factors	Rank																Total
	1	2	3	4	5	6	7	8	9	10	11	12	13	14	15	16	
Drought	33	0	4	10	21	8	2	18	1	1	0	1	1	0	0	1	101
Hailstorm	1	6	6	25	6	5	0	1	0	0	0	0	0	1	33	17	101
Flood	23	5	53	1	0	3	15	0	0	0	0	0	0	0	0	1	101
AD	0	32	1	0	2	3	3	0	17	2	0	1	1	6	18	15	101
IPD	21	25	16	35	2	0	0	0	0	0	0	0	1	0	1	0	101
IFM	0	1	0	0	17	1	1	0	0	8	2	43	18	5	3	2	101
UR	17	2	10	25	42	2	0	0	0	1	1	1	0	0	0	0	101
IW	2	10	5	2	7	63	5	3	3	0	0	0	0	1	0	0	101
TFH	1	0	1	1	1	1	9	14	4	14	0	2	32	19	2	0	101
TFL	0	0	1	0	1	2	5	5	6	3	23	3	3	32	17	0	101
SSDI	0	1	0	1	0	0	1	0	3	4	34	9	18	11	9	10	101
CCP	0	0	0	0	0	0	37	7	13	21	7	4	7	4	1	0	101
CLA	0	0	0	0	0	1	1	13	4	4	22	2	8	2	10	34	101
CFI	0	1	0	0	1	1	18	38	6	6	2	14	6	2	4	2	101
CF	1	18	4	0	1	10	3	2	38	4	8	4	4	2	1	1	101
DGW	2	0	0	1	0	1	1	0	6	33	2	17	2	16	2	18	101
Total	101	101	101	101	101	101	101	101	101	101	101	101	101	101	101	101	1616
Garrett score	86	76	70	65	61	58	55	52	48	45	42	39	35	30	24	14	

Table 9.21 Rankings of factors representing climate change using Garrett ranking: Andhra Pradesh

Factors	Rank																Total score	Average score	Rank
	1	2	3	4	5	6	7	8	9	10	11	12	13	14	15	16			
	Total score																		
Drought	2838	0	280	650	1281	464	110	936	48	45	0	39	35	0	0	14	6740	66.7	4
Hailstorm	86	456	420	1625	366	290	0	52	0	0	0	0	0	30	792	238	4355	43.1	10
Flood	1978	380	3710	65	0	174	825	0	0	0	0	0	0	0	0	14	7146	70.8	2
AD	0	2432	70	0	122	174	165	0	816	90	0	39	35	180	432	210	4765	47.2	8
IPD	1806	1900	1120	2275	122	0	0	0	0	0	0	0	35	0	24	0	7282	72.1	1
IFM	0	76	0	0	1037	58	55	0	0	360	84	1677	630	150	72	28	4227	41.9	11
UR	1462	152	700	1625	2562	116	0	0	0	45	42	39	0	0	0	0	6743	66.8	3
IW	172	760	350	130	427	3654	275	156	144	0	0	0	0	30	0	0	6098	60.4	5
TFH	86	0	70	65	61	58	495	728	192	630	0	78	1120	570	48	0	4201	41.6	12
TFL	0	0	70	0	61	116	275	260	288	135	966	117	105	960	408	0	3761	37.2	13
SSDI	0	76	0	65	0	0	55	0	144	180	1428	351	630	330	216	140	3615	35.8	15
CCP	0	0	0	0	0	0	2035	364	624	945	294	156	245	120	24	0	4807	47.6	7
CLA	0	0	0	0	0	58	55	676	192	180	924	78	280	60	240	476	3219	31.9	16
CFI	0	76	0	0	61	58	990	1976	288	270	84	546	210	60	96	28	4743	47.0	9
CF	86	1368	280	0	61	580	165	104	1824	180	336	156	140	60	24	14	5378	53.2	6
DGW	172	0	0	65	0	58	55	0	288	1485	84	663	70	480	48	252	3720	36.8	14

Table 9.22 Factors influencing farmers perception of climate change—Garrett ranking: Karnataka

Factors	Rank																Total
	1	2	3	4	5	6	7	8	9	10	11	12	13	14	15	16	
Drought	3	2	1	5	0	0	0	0	0	0	0	8	24	12	22	23	100
Hailstorm	3	0	2	1	0	0	0	0	0	0	0	6	12	43	31	2	100
Flood	0	3	3	0	0	0	0	0	3	3	2	1	6	14	42	23	100
AD	0	9	16	0	6	5	3	4	11	9	10	15	6	5	1	0	100
IPD	58	18	4	5	0	2	1	11	1	0	0	0	0	0	0	0	100
IFM	3	0	0	3	0	0	1	3	10	9	21	13	27	5	0	5	100
UR	4	14	10	12	5	7	8	9	13	4	5	3	3	3	0	0	100
IW	2	2	6	14	13	3	17	23	4	6	4	1	0	5	0	0	100
TFH	0	3	5	4	14	13	6	8	9	13	11	11	2	1	0	0	100
TFL	10	0	0	3	3	0	5	2	12	15	10	16	9	8	4	3	100
SSDI	10	8	13	25	17	14	4	1	4	4	0	0	0	0	0	0	100
CCP	0	0	0	3	8	19	18	10	20	10	8	3	1	0	0	0	100
CLA	0	0	1	2	4	10	14	15	4	19	22	3	5	1	0	0	100
CFI	7	27	27	4	20	4	0	0	2	0	6	2	0	1	0	0	100
CF	0	14	12	19	10	23	7	9	2	1	0	3	0	0	0	0	100
DGW	0	0	0	0	0	0	16	5	5	7	1	15	5	2	0	44	100
Total	100	100	100	100	100	100	100	100	100	100	100	100	100	100	100	100	1600
Garret score	86	76	70	65	61	58	55	52	48	45	42	39	35	30	24	14	

Table 9.23 Rankings of factors representing climate change using Garrett ranking: Karnataka

Factors	Rank																Total score	Average score	Rank
	1	2	3	4	5	6	7	8	9	10	11	12	13	14	15	16			
	Total score																		
Drought	2838	0	280	650	1281	464	110	936	48	45	0	39	35	0	0	14	6740	66.7	4
Hailstorm	86	456	420	1625	366	290	0	52	0	0	0	0	0	30	792	238	4355	43.1	10
Flood	1978	380	3710	65	0	174	825	0	0	0	0	0	0	0	0	14	7146	70.8	2
AD	0	2432	70	0	122	174	165	0	816	90	0	39	35	180	432	210	4765	47.2	8
IPD	1806	1900	1120	2275	122	0	0	0	0	0	0	0	35	0	24	0	7282	72.1	1
IFM	0	76	0	0	1037	58	55	0	0	360	84	1677	630	150	72	28	4227	41.9	11
UR	1462	152	700	1625	2562	116	0	0	0	45	42	39	0	0	0	0	6743	66.8	3
IW	172	760	350	130	427	3654	275	156	144	0	0	0	0	30	0	0	6098	60.4	5
TFH	86	0	70	65	61	58	495	728	192	630	0	78	1120	570	48	0	4201	41.6	12
TFL	0	0	70	0	61	116	275	260	288	135	966	117	105	960	408	0	3761	37.2	13
SSDI	0	76	0	65	0	0	55	0	144	180	1428	351	630	330	216	140	3615	35.8	15
CCP	0	0	0	0	0	0	2035	364	624	945	294	156	245	120	24	0	4807	47.6	7
CLA	0	0	0	0	0	58	55	676	192	180	924	78	280	60	240	476	3219	31.9	16
CFI	0	76	0	0	61	58	990	1976	288	270	84	546	210	60	96	28	4743	47.0	9
CF	86	1368	280	0	61	580	165	104	1824	180	336	156	140	60	24	14	5378	53.2	6
DGW	172	0	0	65	0	58	55	0	288	1485	84	663	70	480	48	252	3720	36.8	14

Chapter 10
Economics of Adoption of Rice Production and Management Technologies

K. Palanisami, Krishna Reddy Kakumanu, Udaya Sekhar Nagothu and C. R. Ranganathan

The major factor that may help to mitigate the climate change impacts in the agriculture sector will be introduction of technologies or practices that will help in providing the assured income. List of such technologies or practices adopted by the farmers in rice crop cultivation was obtained from the field surveys from different states and analysed. The key technologies or practices popular among the farmers and useful for upscaling in the states in different regions are discussed in the following sections. Only the states with proven use of the technologies or practices have been included in the detailed analysis.

10.1 Rice Technology Adoption and Profitability

10.1.1 Northern Region

There are many technologies available to farmers for improving their rice cultivation and mitigate the impacts due to climate change. The technologies such as direct sown rice, machine transplanting, system of rice intensification, alternate wetting and drying of rice are some of the practices which are followed up and found profitable (Palanisami et al. 2014). Since details on the spread up of these practices are not available for all the regions, a detailed analysis was done by asking the farmers about the availability, access and adoption of those technologies and also the constraints in adopting them.

The authors would like to thank Dr. K.R. Karunakaran, Prof of Agrl. Economics, Tamilnadu Agrl. University for providing substantial inputs in preparing this chapter without which, the chapter would not have taken the shape that it has now.

K. Palanisami et al., *Climate Change and Future Rice Production in India*, India Studies in Business and Economics,
https://doi.org/10.1007/978-981-13-8363-2_10

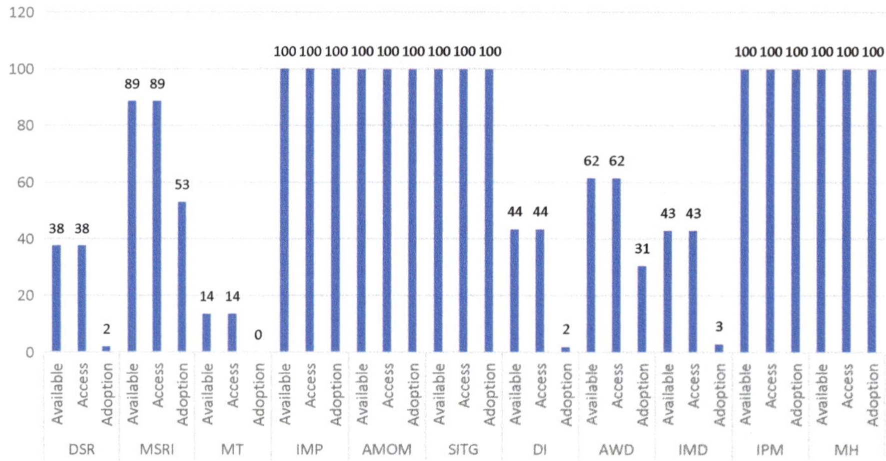

Fig. 10.1 Popularity of rice technologies among farmers in northern region (%). *Note* DSR = direct sown rice; MSRI = modified system of rice intensification; MT = machine transplanting; IMP = improved management practices; AMOM = application of more organic manure; SITG = supplemental irrigation through groundwater; DI = drip irrigation; AWD = alternate wetting and drying; IMD = irrigation at minimum depth; IPM = integrated pest management; MH = machine harvesting

The access, availability and adoption level of 11 rice technologies observed in the region were analysed and presented in Fig. 10.1. Among the 11 rice technologies, improved management practices (IMP), application of more organic manure (AMOM), machine harvesting (MH), integrated pest management (IPM) and supplemental irrigation through groundwater (SITG) are the most popular technologies adopted by all the sample farmers. The least adopted technologies are machine transplanting (MT), direct sown rice (DSR), drip irrigation (DI) and irrigation at minimum depth (IMD).

10.1.1.1 Costs and Benefits of Technology Adoption—Punjab

In order to get an idea on the economics of technology adoption at state level, the farm-level cost of cultivation (COC), cost of technology, yield, gross return and net return were computed and are presented in Table 10.1

The table shows that an increase in net income over the conventional method (CM) was higher with alternate wetting and drying (AWD) (Rs. 19,142/ha) followed by MSRI (Rs. 17,882/ha) and improved management practices (IMP) (Rs. 15,342/ha). The table also highlights that in terms of cost of production of rice per quintal (qtl), it is invariably less under all the technologies/practices compared to the conventional method, thus providing incentives to the farmers for technology adoption. Nonetheless, the cost of technology is increased from Rs. 2000 to 7000/ha by adopting the

Table 10.1 Economics of rice technologies—Punjab

Technology	Cost of cultivation (COC) (Rs./ha)	Cost of Tech. (Rs./ha)	Yield (t/ha)	Gross Return (Rs./ha)	Net Return (NI) (Rs./ha)	Increase in NI over CM (Rs./ha)	Cost of production (COP) (Rs./qtl)
CM	35,392		4.9	62,750	27,358		722
DSR	33,200	−2192	4.8	62,300	29,100	1742	692
MSRI	40,500	5108	6.6	85,740	45,240	17,882	614
IMP	38,600	3208	6.2	81,300	42,700	15,342	623
AMOM	41,400	6008	6.1	79,650	38,250	10,892	679
SITG	42,640	7248	6.4	83,100	40,460	13,102	666
AWD	37,400	2008	6.5	83,900	46,500	19,142	575

Note CM conventional method; *DSR* direct sown rice; *MSRI* modified system of rice intensification; *MT* machine transplanting; *IMP* improved management practices; *AMOM* application of more organic manure; *SITG* supplemental irrigation through groundwater; *AWD* alternate wetting and drying

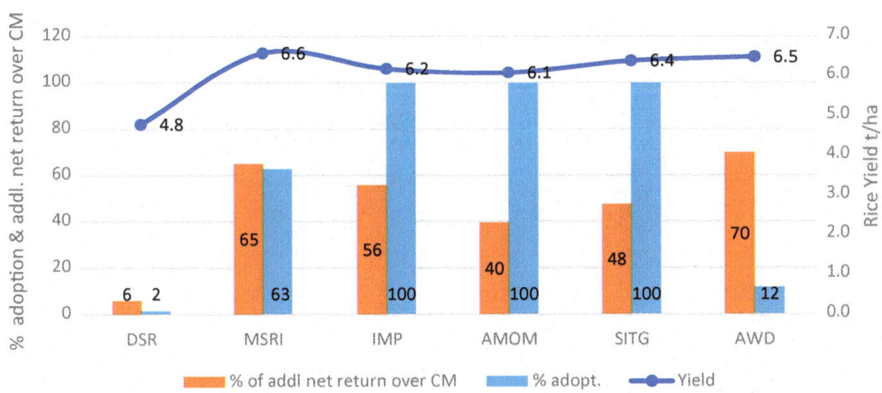

Fig. 10.2 Technology-wise % adoption, % addl.net income over CM and rice yield in Punjab

technologies. But, the cost of technology for adopting DSR is reduced (Rs. 2192/ha) due to direct seeding by withdrawing labour for transplantation. Figure 10.2 illustrates that in Punjab, adoption percentage is more for improved management practices (IMP), application of more organic manure (AMOM) and supplemental irrigation through groundwater (SITG) with 100%. Among the technologies, though all the farmers have adopted these technologies, the highest productivity was found to be in MSRI (6.6 t/ha) followed by AWD (6.5 t/ha), SITG (6.4 t/ha), IMP (6.2 t/ha) and AMOM (6.1 t/ha). However, AWD realized 6.5 t/ha yield gaining 70% higher additional returns over CM which is presently followed by 12% of the rice farmers.

10.1.1.2 Costs and Benefits of Technology Adoption—Haryana

Table 10.2 gives summary of the cost and benefits for different practices followed by farmers in Haryana. In Haryana also, AWD and MSRI are the two most profitable technologies among seven technologies followed by the rice farmers.

In the case of Haryana, an increase in net income over the CM was higher with AWD (Rs. 13,661/ha) followed by MSRI (Rs. 8393/ha). All other technologies are yielding only marginal increase over CM, and this might be due to comparatively lesser yield than in Punjab. In the case of cost of production of rice per quintal, in most cases, it is marginally higher than CM, thus implying that some of the technologies are not financially attractive to adopt and this might be the reason for their low level of adoption (see Fig. 10.3).

The IMP, AMOM and SITG practices have 100% adoption, but the DI and IMD are adopted by 2–6% only due to the higher cost. The yield benefits are also less

Table 10.2 Economics of rice technologies—Haryana

Technology	Cost of cultivation (COC) (Rs./ha)	Cost of Tech. (Rs./ha)	Yield (t/ha)	Gross return (Rs./ha)	Net return (NI) (Rs./ha)	Increase in NI over CM (Rs./ha)	Cost of production (COP) (Rs./qtl)
CM	31,681		4.8	59,520	27,839		660
MSRI	38,168	6487	6	74,400	36,232	8393	636
IMP	40,791	9110	5.5	68,660	27,869	30	742
AMOM	36,705	5024	5.4	66,960	30,255	2416	680
SITG	39,850	8169	5.7	70,680	30,830	2991	699
AWD	34,140	2459	6.1	75,640	41,500	13,661	560
DI	43,540	11,859	5.9	73,340	29,800	1961	738
IMD	32,980	1299	4.9	60,960	27,980	141	673

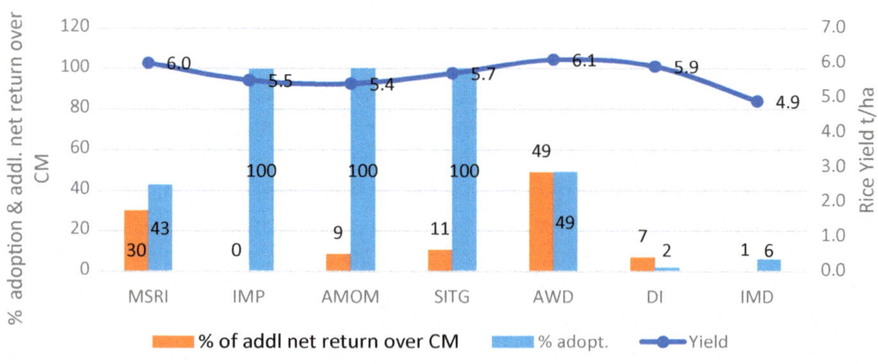

Fig. 10.3 Technology-wise % adoption, % addl.net return over CM and rice yield in Haryana

Table 10.3 Economics of rice technologies in UP

Technology	Cost of cultivation (COC) (Rs./ha)	Cost of Tech. (Rs./ha)	Yield (t/ha)	Gross Return (Rs./ha)	Net Return (NI) (Rs./ha)	Increase in NI over CM (Rs./ha)	Cost of production (COP) (Rs./qtl)
CM	29,138		4.8	60,240	31,102		607
DSR	31,200	2062	5	63,250	32,050	948	624
IMP	36,780	7642	5.4	68,310	31,530	428	681
AMOM	32,705	3567	5.1	64,515	31,810	708	641
SITG	31,850	2712	5.3	67,045	35,195	4093	601
MSRI	33,450	4312	5.6	68,750	35,300	4198	567
MH	32,980	3842	5.1	64,120	31,140	38	673

for IMD. Though DI has higher yield (5.9 t/ha) and water use efficiency, the cost of production (Rs. 738/qtl) is comparatively high, resulting in low adoption level.

10.1.1.3 Costs and Benefits of Technology Adoption—Uttar Pradesh

In case of Uttar Pradesh (UP), increase in net income over the conventional method (CM) was higher with MSRI (Rs. 4198/ha) followed by SITG (Rs. 4093/ha). As in Haryana, all other technologies are yielding only marginal increase over the CM due to comparatively lesser yield. The average yield increase under the technologies is only about 9% over the CM compared to about 14% increase in average cost of cultivation. In most cases, the cost of production per quintal is marginally higher than the CM, thus warranting efforts to minimize the cost of cultivation with more emphasis on cost-effective adoption mechanisms including field level training. Interestingly, the cost of production with MSRI is less than the CM due to higher yield (5.6 t/ha) (Table 10.3). However, the adoption rate under MSRI and DSR is less, which is about 2% (Fig. 10.4). The cost of technology in DSR is also higher in the state due to the increased cost of herbicides and fertilizers.

10.1.2 Eastern Region

Among the ten rice technologies identified in the region, MH, IMP, SITG and IPM are the most popular technologies. Though the availability of these technologies to the farmers is reported to be high, viz. 45, 81, 48 and 23%, the access to these technologies by the farmers was 40, 81, 32 and 40%, respectively (Fig. 10.5). In the case of adoption, it was only 34, 64, 26 and 15%, whereas MSRI and MT were adopted

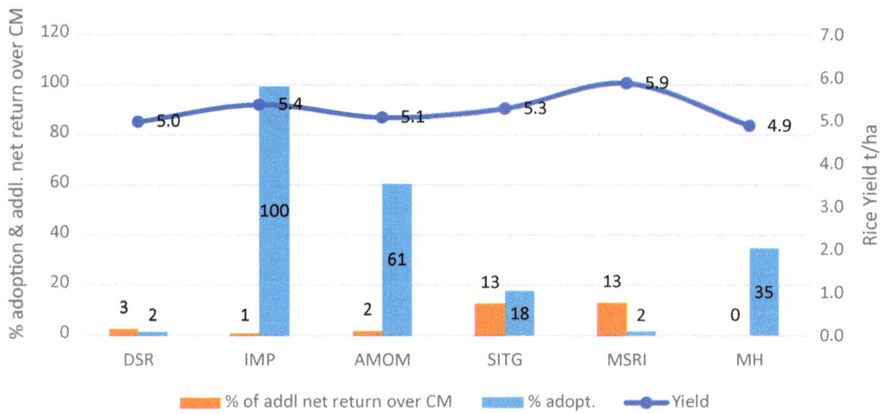

Fig. 10.4 Technology-wise % adoption, % addl.net Income over CM and rice yield in Uttar Pradesh

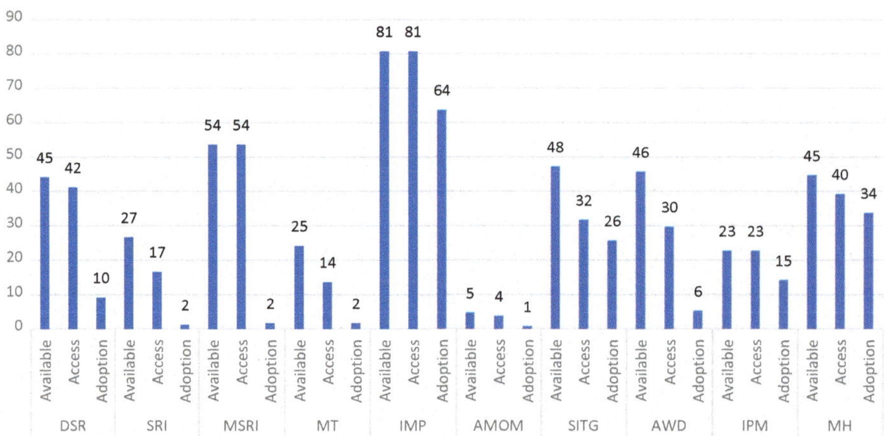

Fig. 10.5 Popularity of rice technologies among farmers in eastern region (%)

by 1.5 and 2% farmers, respectively, indicating the need for initiating upscaling measures.

10.1.2.1 Costs and Benefits of Technology Adoption—Odisha

Table 10.4 shows that in Odisha, the average cost of cultivation varies from Rs. 30,650/ha under CM to Rs. 34,250/ha with DSR. The net returns (NI) are higher for IMP and MSRI practices. Cost of production in all the adopted technologies is slightly lower than the CM. Among these technologies, the cost of production under MSRI is comparatively lower (Rs. 551/qtl) followed by IMP (Rs. 584/qtl) and MH and DSR both at Rs. 591/qtl.

Table 10.4 Economics of rice technologies in Odisha

Technology	Cost of cultivation (COC) (Rs./ha)	Cost of Tech. (Rs./ha)	Yield (t/ha)	Gross Return (Rs./ha)	Net Return (NI) (Rs./ha)	Increase in NI over CM (Rs./ha)	Cost of production (COP) (Rs./qtl)
CM	30,650		4.8	52,800	22,150		634
IMP	33,311	2661	5.7	65,550	32,239	10,089	584
DSR	34,250	3600	5.8	63,800	29,550	7400	591
MSRI	32,490	1840	5.9	64,900	32,410	10,260	551
MT	33,540	2890	5.5	60,500	26,960	4810	610
MH	33,100	2450	5.6	61,600	28,500	6350	591

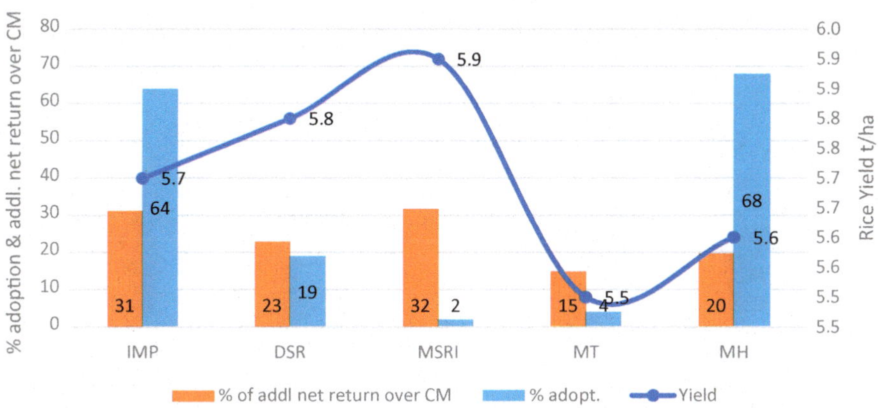

Fig. 10.6 Technology-wise % adoption, % addl.net return over CM and rice yield in Odisha

MH and IMP are adopted by more than half of the farmers. The DSR, MSRI and MT, which are very important to address the climate change impacts and improve water use efficiency, are adopted by only 19, 2, and 4%, respectively (Fig. 10.6). The yields are also higher in the MSRI (5.9 t/ha) and DSR (5.8 t/ha) when compared to the other adoption practices.

10.1.2.2 Costs and Benefits of Technology Adoption—West Bengal

The details of the cost and return of technology adoption in West Bengal are presented in Table 10.5. The average cost of cultivation varies from Rs. 27,258/ha under CM to Rs. 33,800/ha with AMOM. The rice yield is uniformly higher with the technologies, showing the impact of technology adoption in boosting the rice yield. Increase in net income over the CM was higher with SRI, MSRI, AWD and IPM. In terms of cost of production, most of the technologies yielded varying results from CM implying

Table 10.5 Economics of rice technologies—West Bengal

Technology	Cost of cultivation (COC) (Rs./ha)	Cost of Tech. (Rs./ha)	Yield (t/ha)	Gross Return (Rs./ha)	Net Return (NI) (Rs./ha)	Increase in NI over CM (Rs./ha)	Cost of production (COP) (Rs./qtl)
CM	27,258		4.9	54,880	27,622		556
SRI	33,420	6162	6.2	70,200	36,780	9158	539
IMP	31,456	4198	5.7	64,500	33,044	5422	552
MSRI	32,490	5232	6.1	68,650	36,160	8538	533
AMOM	33,800	6542	5.4	61,620	27,820	198	626
SITG	33,438	6180	5.8	64,580	31,142	3520	577
AWD	32,450	5192	6.0	67,200	34,750	7128	541
IPM	31,840	4582	5.9	66,080	34,240	6618	540

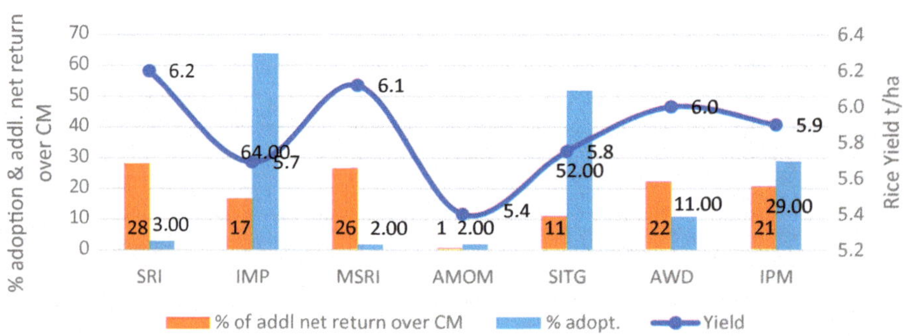

Fig. 10.7 Technology-wise % adoption, % addl.net return over CM and rice yield in West Bengal

that technology-induced cost components need to be examined in detail so that cost minimization efforts can be targeted.

IMP and SITG are adopted more than 50% in West Bengal. The SRI, MSRI and AWD practices which are high yielding (6.2, 6.1 and 6.0 t/ha, respectively) in the state are less adopted possibly due to lack of proper awareness (Fig. 10.7).

10.1.3 Western Region

The available, access and adoption level of different rice technologies in Gujarat are presented in Fig. 10.8. Applying more organic manure (AMOM), MSRI and SITG are the three prominent technologies widely available to and accessed by the Gujarat rice farmers. However, the actual adoption rates were 48, 35 and 28%, respectively.

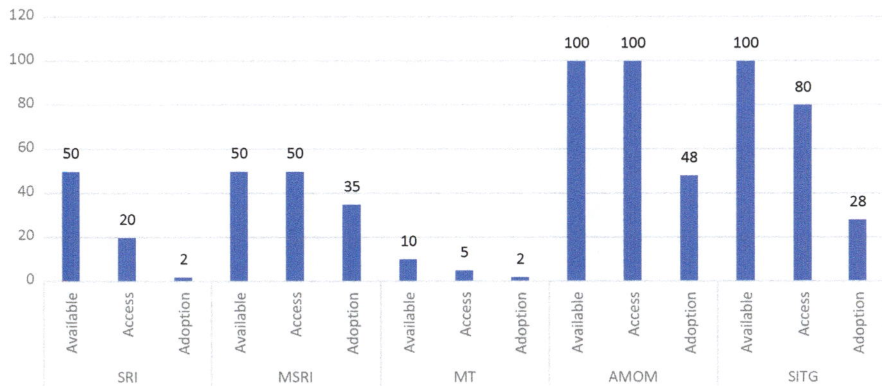

Fig. 10.8 Popularity of rice technologies among farmers in western (%)

Table 10.6 Economics of rice technologies in Gujarat

Technology	Cost of cultivation (COC) (Rs./ha)	Cost of Tech. (Rs./ha)	Yield (t/ha)	Gross Return (Rs./ha)	Net Return (NI) (Rs./ha)	Increase in NI over CM (Rs./ha)	Cost of production (COP) (Rs./qtl)
CM	30,075		5.1	62,317	32,242		590
SRI	37,680	7605	6.4	77,658	39,978	7736	589
MSRI	36,530	6455	6.2	75,890	39,360	7118	589
MT	30,870	795	5.7	71,475	40,605	8363	542
AMOM	34,890	4815	5.6	69,364	34,474	2232	623
SITG	37,450	7375	6.1	75,640	38,190	5948	614

MT is another important labour-saving rice technology which is adopted by only 2% of the rice farmers in Gujarat.

The costs and benefits of technology adoption are presented in Table 10.6. The average cost of cultivation varies from Rs. 30,075/ha under CM to Rs. 37,680/ha with SRI. In the case of increased net income over the conventional methods, all the adopted technologies resulted in higher net income though the magnitude varies across the technologies. Among the technologies, the cost of production under MT is comparatively lower (Rs. 542/qtl) than other technologies. However, the cost of production has not much varied among the technologies which are almost similar to the cost under CM.

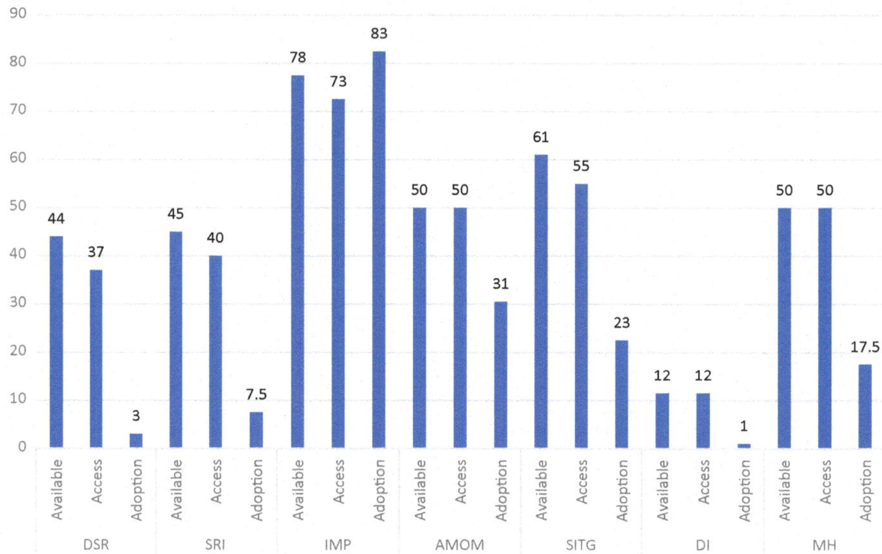

Fig. 10.9 Popularity of rice technologies among farmers in central region (%)

10.1.4 Central Region

The types of technologies available, access and adoption level in the central region were studied in detail and are presented in Fig. 10.9. Among the identified rice technologies, seven technologies were available and reported. Among them, IMP, AMOM and SITG are the three popular technologies whose adoption rates were 83, 31 and 17.5%, respectively. It is important to examine why some of the technologies popular in other regions such as SRI, MSRI are not being followed in the region. Cross-learning of the technology adoption process will help identify such adoption constraints.

10.1.4.1 Costs and Benefits of Technology Adoption—Madhya Pradesh

The cost and additional return realized under four major rice technologies in MP were analysed, and the results are presented in Table 10.7.

The table shows that the average cost of cultivation varies from Rs. 31,803/ha under CM to Rs. 37,890/ha with SRI (i.e. 19% higher). In terms of increased net income, except DSR, all other technologies showed a similar increase. The cost of technology is negative for DSR, implying that compared to conventional (transplanting) method, DSR resulted in lesser cost, and hence, negative and in practice we can say DSR involves no additional cost towards technology and this is also reflected in the reduction in the total cost of cultivation compared to conventional method

Table 10.7 Economics of rice technologies in Madhya Pradesh

Technology	Cost of cultivation (COC) (Rs./ha)	Cost of Tech. (Rs./ha)	Yield (t/ha)	Gross Return (Rs./ha)	Net Return (NI) (Rs./ha)	Increase in NI over CM (Rs./ha)	Cost of production (COP) (Rs./qtl)
CM	31,803		4.8	54,000	22,197		663
DSR	29,100	−2703	4.9	54,860	25,760	3563	594
IMP	32,780	977	5.4	62,300	29,520	7323	607
SITG	35,650	3847	5.9	65,480	29,830	7633	604
SRI	37,890	6087	6.2	69,560	31,670	9473	611

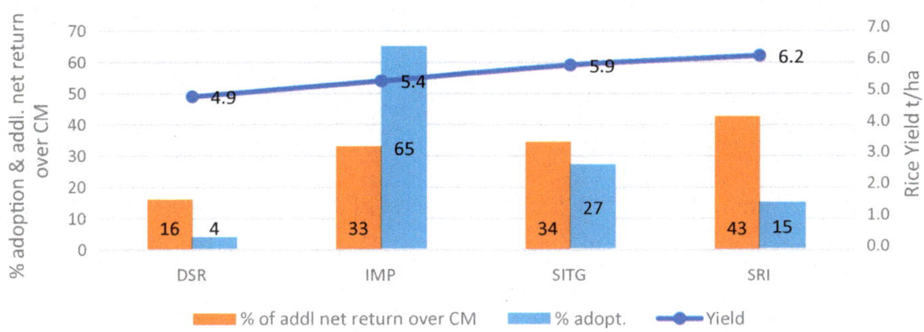

Fig. 10.10 Technology-wise % adoption, % addl.net return over CM and rice yield in Madhya Pradesh

(Kakumanu et al. 2018). The cost of production under all the adopted technologies is lower than the conventional method, and among them, DSR is seemingly much lower (Rs. 594/qtl) due to less labour use and overcoming transplanting operations.

However, IMP adoption rate is higher (65%) followed by SITG (27%) and SRI (15%) with higher rice productivity in SRI (6.2 t/h) and SITG (5.9 t/ha) (Fig. 10.10). This implies the need for initiating more awareness programs and field demonstrations to upscale these technologies as their cost of production is also less compared to CM.

10.1.5 Southern Region

Among the rice technologies, machine harvesting (MH), improved management practices (IMP), supplemental irrigation (SIGT) and integrated pest management (IPM) are more popular. The percentage adoption level of these technologies varied from 31 to 71 (Fig. 10.11). It is interesting to observe that though these technologies

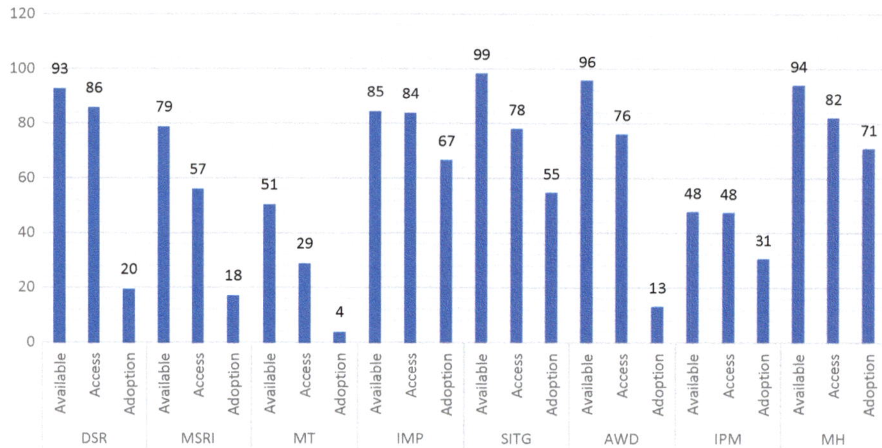

Fig. 10.11 Popularity of rice technologies among farmers in southern region (%)

are known for their higher yield, their lower level of adoption indicates the need for examining the constraints in their adoption.

10.1.5.1 Costs and Benefits of Technology Adoption—Andhra Pradesh

In the case of economics of technology adoption, the average cost of cultivation varies from Rs. 37,830/ha under CM to Rs. 45,983/ha with SITG in Andhra Pradesh (Table 10.8). In all the cases, the cost of cultivation has been higher by 3% (for DSR) to 21% (for SITG) indicating the importance of the cost in technology introduction and adoption. In case of DSR, a recent study shows that there is a decrease in cost of cultivation compared to CM (Kakumanu et al. 2018). In addition, the study also explained the benefits of DSR and the role in mitigating climate change impacts.

In the case of increased net income, MSRI, AWD, IMP and IPM have comparatively higher income than SSD and MH. The cost of production under AWD is comparatively lower (Rs. 653/qtl) than SSD (Rs. 826/qtl). However, as observed in other regions, the cost of production with technology and without technology is not much significant, and this might due to different levels of increase in cost of cultivation and net income.

AWD and MT are also practised in the state but to a very less extent (Fig. 10.12). It is mainly due to the non-availability of machinery for timely transplantation and lack of on- and off-systems in canal operation. IMP and IPM are also ighly practised in the state as there is a need to control increasing pest and diseases to the changing climate.

Table 10.8 Economics of rice technologies—Andhra Pradesh

Technology	Cost of cultivation (COC) (Rs./ha)	Cost of Tech. (Rs./ha)	Yield (t/ha)	Gross Return (Rs./ha)	Net Return (NI) (Rs./ha)	Increase in NI over CM (Rs./ha)	Cost of production (COP) (Rs./qtl)
CM	37,830		4.8	59,040	21,210		788
MSRI	43,284	5454	6.6	75,640	32,356	11,146	656
AWD	39,200	1370	6.0	72,860	33,660	12,450	653
SITG	45,983	8153	6.2	73,450	27,467	6257	742
DSR	39,130	1300	5.1	66,400	27,270	6060	767
SSD	43,765	5935	5.3	65,190	21,425	215	826
MT	39,745	1915	5.4	66,420	26,675	5465	736
IMP	39,560	1730	5.9	72,570	33,010	11,800	671
IPM	41,240	3410	6.1	75,030	33,790	12,580	676
MH	39,870	2040	5.1	62,730	22,860	1650	782

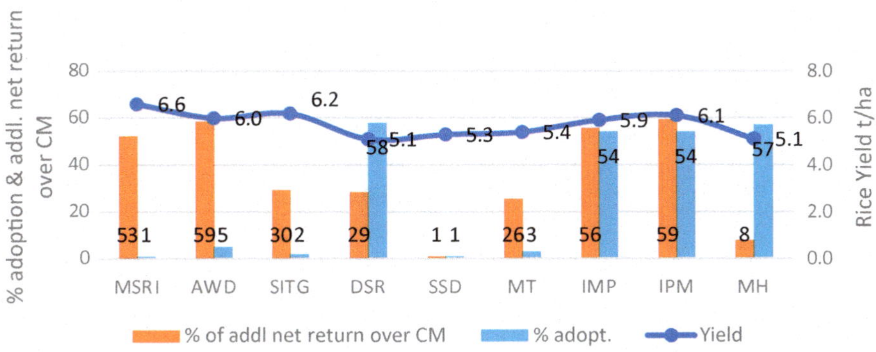

Fig. 10.12 Technology-wise % adoption, % addl.net return over CM and rice yield in Andhra Pradesh

10.1.5.2 Costs and Benefits of Technology Adoption—Tamil Nadu

In case of Tamil Nadu, the average cost of cultivation varies from Rs. 29,740/ha under CM to Rs. 37,680/ha with SITG. One interesting aspect is that yield increase is significantly higher with technology adoption where the average yield under technologies as a whole is about 6.3 t/ha compared to 5.1/ha with conventional method. Increase in net income is higher with IMP (Rs. 12,015/ha) which is about 38% higher than the CM, whereas the cost of production under AWD is lower at Rs. 526/qtl followed by IMP and MH.

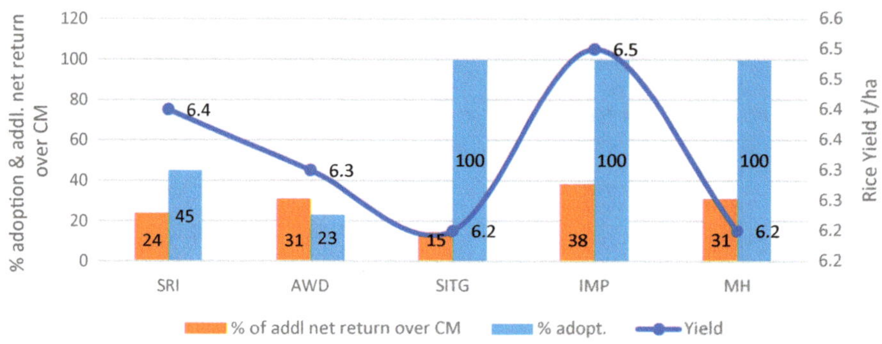

Fig. 10.13 Technology-wise % adoption, % addl.net return over CM and rice yield in Tamil Nadu

Table 10.9 Economics of rice technologies—Tamil Nadu

Technology	Cost of cultivation (COC) (Rs./ha)	Cost of Tech. (Rs./ha)	Yield (t/ha)	Gross Return (Rs./ha)	Net Return (NI) (Rs./ha)	Increase in NI over CM (Rs./ha)	Cost of production (COP) (Rs./qtl)
CM	29,740		5.1	61,245	31,505		583
SRI	36,580	6840	6.4	75,520	38,940	7435	572
AWD	33,120	3380	6.3	74,320	41,200	9695	526
SITG	37,680	7940	6.2	73,755	36,075	4570	608
IMP	35,130	5390	6.5	78,650	43,520	12,015	540
MH	33,765	4025	6.2	75,020	41,255	9750	545

The SITG, IMP and MH are fully adopted (100%) by the sample farmers (Fig. 10.13). Yields also have comparative advantage in technology adoption (1.1–1.3 t/ha). Cost of production is higher for SITG when compared to CM. All other four practices are generally showing comparatively lower cost than the CM (Table 10.9).

10.1.5.3 Costs and Benefits of Technology Adoption—Karnataka

In the case of Karnataka, the average cost of cultivation varies from Rs. 34,830/ha under conventional method to Rs. 45,983/ha with SITG (Table 10.10). Regarding the technology cost, the supplemental irrigation costs are more due to possible water scarcity, but the yield is higher (6.2 t/ha). In the case of increased net income over conventional method, only SRI resulted in higher income (Rs. 10,396/ha) and the

Table 10.10 Economics of rice technologies—Karnataka

Technology	Cost of cultivation (COC) (Rs./ha)	Cost of Tech. (Rs./ha)	Yield (t/ha)	Gross Return (Rs./ha)	Net Return (NI) (Rs./ha)	Increase in NI over CM (Rs./ha)	Cost of production (COP) (Rs./qtl)
CM	34,830		5.3	76,850	42,020		657
DSR	35,130	300	5.4	78,300	43,170	1150	651
SRI	43,284	8454	6.6	95,700	52,416	10,396	656
IMP	36,500	7950	5.4	78,600	42,100	80	676
SITG	45,983	11,153	6.2	89,900	43,917	1897	742
AWD	37,130	4300	5.6	81,200	44,070	2050	663
IPM	41,050	6220	5.8	84,100	43,050	1030	708
MH	35,767	8935	5.5	79,750	43,983	1963	650
MT	36,765	1935	5.6	81,200	44,435	2415	657

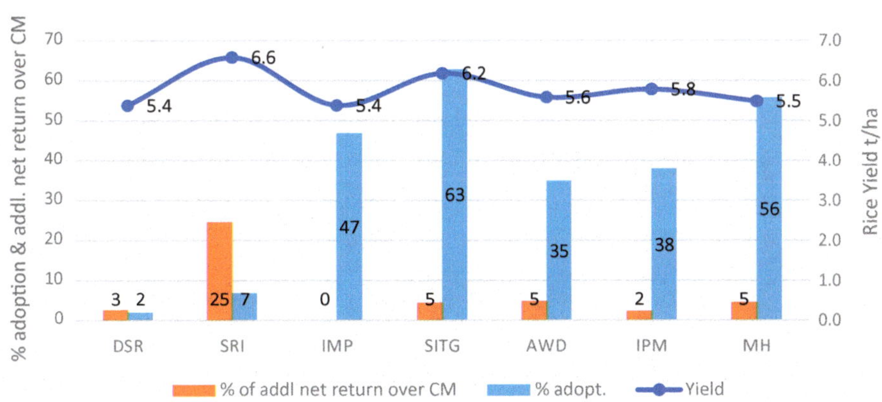

Fig. 10.14 Technology-wise % adoption, % addl.net return over CM and rice yield in Karnataka

least was observed under IMP and this might be due to lower rice yield. Higher cost of cultivation and marginal yield increase with technologies resulted in only marginal increase in net income over the conventional method, thus warranting strategies to increase the yield (Fig. 10.14).

Adoption of SITG, MH and IMP practices are high in Karnataka state. The percentage of additional net returns over CM is more for SRI followed by AWD, SITG and MH accounting for 5% each.

Table 10.11 Increase in cost of cultivation with technologies over the conventional method (%)

Technology	Northern			Western	Eastern		Central	Southern			Average
	PU	HA	UP	GU	OD	WB	MP	AP	TN	KA	
AMOM	16.98	15.86	12.24	16.01		24					17.02
AWD	5.67	7.76				19.05		3.62	6.6	11.37	9.01
DI		37.43									37.43
DSR	−6.19		7.08		11.75		−8.5	3.44	0.86		1.41
IMD		4.1									4.10
IMP	9.06	28.76	26.23		8.68	15.4	3.07	4.57	4.79	18.12	13.19
IPM						16.81		9.01	17.86		14.56
MH			13.19		7.99			5.39	2.69	13.53	8.56
MSRI	14.43	20.48	14.8	21.46	6	19.19		14.42			15.83
MT				2.64	9.43			5.06	5.56		5.67
SITG	20.48	25.79	9.31	24.52		22.67	12.1	21.55	32.02	26.7	21.68
SRI				25.29		22.61	19.14		24.27	23	22.86
SSD								15.68			15.68

Note DSR direct sown rice; *MSRI* modified system of rice intensification; *MT* machine transplanting; *IMP* improved management practices; *AMOM* application of more organic manure; *SITG* supplemental irrigation through groundwater; *DI* drip irrigation; *AWD* alternate wetting and drying; *IMD* irrigation at minimum depth; *IPM* integrated pest management; *MH* machine harvesting *SSD* subsurface drainage

10.2 Comparison of Cost and Yield Attributes of Rice Due to Technology Adoption

It is interesting to note that in most cases discussed earlier, the cost of production per quintal (qtl) was almost same with and without technology adoption. This is possibly due to higher cost of cultivation without corresponding increase in yield. Otherwise, both cost and yield did not change much due to technology adoption. In order to get an idea on this, a comparison of the cost of cultivation under different technologies across regions has been made (Table 10.11). It is seen that invariably in all the regions, the cost of cultivation of rice with the technology adoption has increased from negligible percentage with DSR to as high as 37% with DI. Average increase in cost shows 37% for DI followed by 23% with SRI, 22% with SITG, 17% with AMOM, etc. The high cost under DI might be due to its initial high cost of drip irrigation system. However, this technology has been in its initial stage of adoption under government/research support programs. The cost of cultivation for DSR in Punjab and MP states is lesser than CM due to saving in labour cost.

In the case of yield increase, except AMOM, DI, SSD, other technologies exhibited increased yields than the cost increment (Table 10.12). In the case of SITG, there is no significant difference. Since the difference in the cost and yield increase is varying across regions and technologies as seen in Fig. 10.15, technology-focused efforts are needed to increase the yield and reduce the cost.

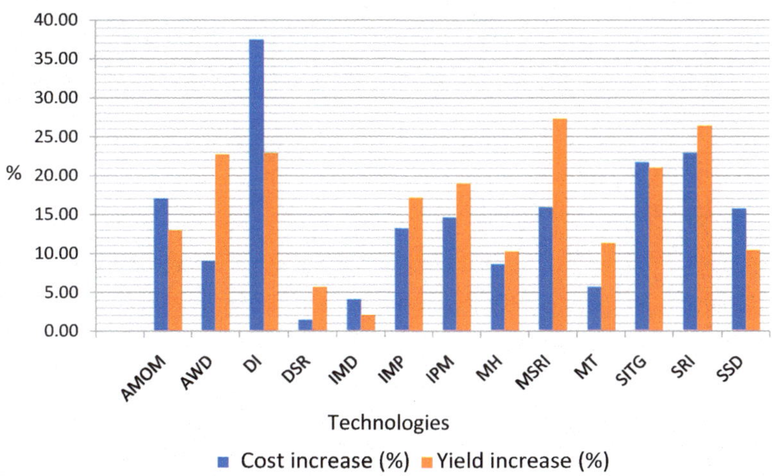

Fig. 10.15 Technology-wise cost and yield increase

Table 10.12 Increase rice yield with technologies over the conventional method (%)

Technology	Northern			Western	Eastern		Central	Southern			Average
	PU	HA	UP	GU	OD	WB	MP	AP	TN	KA	
AMOM	24.49	12.5	7.14	10.45		10.2					12.96
AWD	32.65	27.08				22.45		25	23.53	5.66	22.73
DI		22.92									22.92
DSR	−2.04		5.04		20.83		2.08	6.25		1.89	5.68
IMD		2.08									2.08
IMP	26.53	14.58	13.45		18.75	16.16	12.5	22.92	27.45	1.89	17.14
IPM						20.41		27.08		9.43	18.97
MH			2.94		16.67			6.25	21.57	3.77	10.24
MSRI	34.69	25	23.95	22.29	22.92	24.9		37.5			27.32
MT				12.43	14.58			12.5		5.7	11.30
SITG	30.61	18.75	11.34	20.32		17.37	22.92	29.17	21.57	16.98	21.00
SRI				26.23		26.53	29.17		25.49	24.53	26.39
SSD								10.42			10.42

Note *DSR* direct sown rice; *MSRI* modified system of rice intensification; *MT* machine transplanting; *IMP* improved management practices; *AMOM* application of more organic manure; *SITG* supplemental irrigation through groundwater; *DI* drip irrigation; *AWD* alternate wetting and drying; *IMD* irrigation at minimum depth; *IPM* integrated pest management; *MH* machine harvesting *SSD* subsurface drainage

10.3 Constraints in Technology Adoption

The major constraints faced by the farmers are lack of skills in adopting the technologies and the cost involved in actually getting these technologies in the field. A recent study indicated that transaction cost associated with the technology adoption is higher and this discouraged the farmers in their timely adoption (Palanisami et al. 2015a). Another study on technology adoption had indicated that besides timely water supplies, technologies not matching with the local situation, high capital or initial cost of some of the technologies such as drip and sprinkler, maintenance problems, etc. are the major constraints (Table 10.13) (Palanisami et al. 2015b; GOI 2008).

Table 10.13 Major constraints in technology adoption

State	Total no. of farmers	High capi-tal/initial cost	Too technical to adopt	Lack of skills	Maintenance problem	Need skilled labour	Not matching with local situation
Northern region							
Haryana	5 (100)	–	–	–	2 (40)	–	3 (60)
Punjab	94 (100)	28 (30)	7 (8)	1 (1)	4 (4)	16 (17)	38 (39)
Uttar Pradesh	144 (100)	40 (28)	27 (19)	21 (15)	34 (24)	14 (10)	8 (6)
Western region							
Gujarat	212 (100)	32 (15)	23 (11)	46 (22)	51 (24)	12 (6)	48 (22)
Eastern region							
Odisha	8 (100)	–	–	8 (100)	–	–	–
West Bengal	464 (100)	48 (10)	35 (8)	11 (2)	100 (22)	21 (5)	249 (53)
Central region							
Madhya Pradesh	18 (100)	–	–	8 (44)	–	–	10 (56)
Southern region							
Tamil Nadu	90 (100)	29 (32)	16 (18)	19 (21)	18 (20)	8 (9)	0
Karnataka	480 (100)	197 (41)	21 (5)	75 (16)	14 (3)	115 (24)	58 (12)
Andhra Pradesh	192 (100)	70 (36)	9 (5)	24 (13)	22 (11)	50 (26)	17 (9)

Source Palanisami et al. 2015; Field surveys of FPARP (GOI 2008)
Figures in parentheses indicate percentage to total number of farmers

10.4 Summary

In the northern region, IMP, AMOM, SITG and MH are the major technologies adopted. In the central region SRI, SITG, DSR and IMP are the major technologies adopted with higher net returns from DSR, IMP and SITG. The eastern region is familiar with IMP, DSR, MSRI, MT and MH technologies and among them; MSRI, IMP and DSR realized higher net income. In the western region, SRI, MSRI, SITG, MT and AMOM are the technologies noted for their higher adoption level. In the southern region, MSRI, AWD, SITG, DSR, SSD and MT are observed for making high net income compared to traditional practices. Interestingly, the cost of production of rice with technology adoption has not resulted in cost reduction compared to conventional method mainly due to more increase in cost of cultivation of rice than yield increase thus highlighting need for introducing measures to increase the rice productivity with reduced cultivation costs. This may be possible when the adoption level as a whole is increased in all the regions, as the present adoption level of the technologies is only from 5 to 30%. The major constraints faced by the farmers in technology adoption include lack of technical skills, high initial cost of the technology, untimely canal water supplies and high transaction cost in actually getting these technologies to the field. Therefore, region-specific technology promotion programs such as awareness and capacity building could help increase the adoption of these technologies.

Chapter 11
Climate Change and Rice Production in India: A Way Forward

K. Palanisami, Krishna Reddy Kakumanu, Udaya Sekhar Nagothu and C. R. Ranganathan

11.1 Introduction

It is expected that climate change and its variability could have serious impact on agriculture and water sectors and eventually on the food security and livelihoods of a large section of the rural population in developing countries (IPCC AR4 2007; IPCC 2018). The climate change, in the long run, may lead to new combinations of changes with regard to atmospheric constituents, temperature, solar radiation and pests, diseases and weeds. Consequently, in the Indian subcontinent, such expected changes in climate would alter regional agricultural systems thereby a reduction in rice production, thus leading to food insecurity.

The results based on region-wise analysis evinced that climate change has an impact on rice productivity and production. Thus, there is an increasing concern among farmers, researchers and policymakers due to potential impacts, the future challenges and ways and means to deal with such challenges. The rice, being the major staple food crop in India, the expected yield reductions due to climate change are seemingly critical. By some estimates, over a quarter of all rice production in a majority of the river basins in India may be lost due to climate change and its variability leading to extreme weather consequences like frequent floods and droughts (Palanisami et al. 2014). Climate modellers are using innovative approaches to predict possible climate change and providing inputs to plan future strategies. This has to be seriously considered by the governments to invest in preparing the farming sector to cope with the impacts and sustain rice production and income of rice farmers.

K. Palanisami et al., *Climate Change and Future Rice Production in India*, India Studies in Business and Economics,
https://doi.org/10.1007/978-981-13-8363-2_11

11.2 Overview of Climate Change Impact on Rice Production in India

Currently, the area under rice in the country is about 44 million hectares. The total area under rice in 13 rice growing states selected in this study is approximately 38.44 million hectares constituting 87% of the total area and 89% of the total rice production in the country. As discussed in Chap. 2, the overall growth of rice area during the four decades was 0.4%, the growth in yield was 1.95% and the growth in rice production was 2.36%. Except for a decline in growth rates in states like Bihar, Chhattisgarh, Madhya Pradesh and Tamil Nadu, the overall trends in the area, productivity and production showed positive growth rates over the years. However, when climate change scenarios were considered, future rice production in different regions of India has shown varying levels of rice production as documented in Table 11.1 and Fig. 11.1.

In the northern region, as per RCP4.5 (medium emission) and RCP8.5 (high emission) scenarios, there would be a substantial reduction in rice yield during 2021–2050 (mid-century) and 2071–2100 (end-century) periods. Under both the scenarios, the percentage reduction in the later period will be more. The changes in mean yield show a variation from 6.9 to 12.7% as per RCP4.5 scenario and 8.5–21.4% as per RCP8.5 scenario. Yield variability as estimated by RCP8.5 scenario is more than that predicted by RCP4.5 scenario for both periods. The first scenario predicts an increase of 28.7% yield variability during 2021–2050, while the second scenario predicts an increase of 47% for the same period. For the second period, RCP4.5 predicts a small

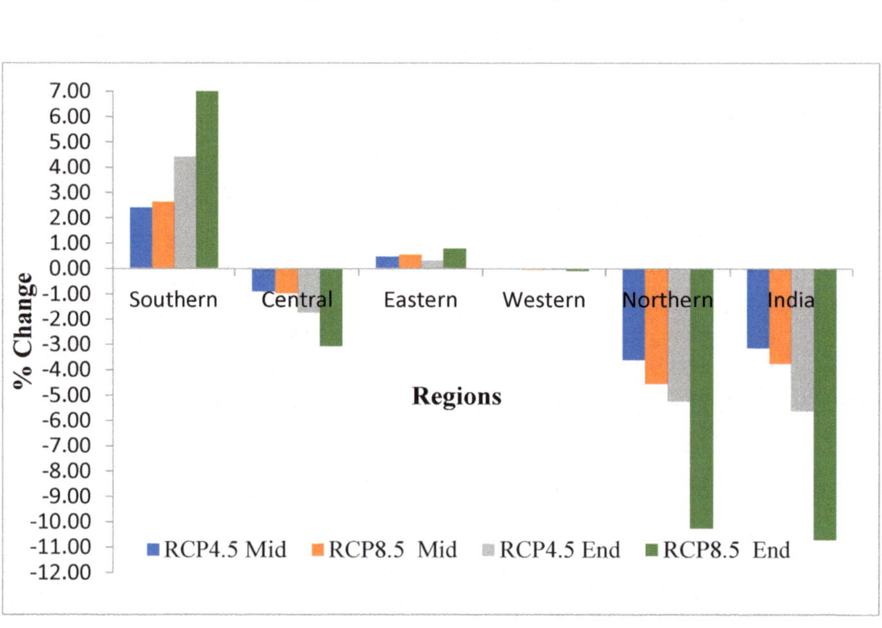

Fig. 11.1 Region-wise impact of climate change on rice yield under different scenarios

Table 11.1 Impact of climate change on rice yield and production in the future, India

Regions	Current area (Million ha)	Current yield (kg/ha)	Current production (Million tonnes)	Projected area (Million ha)	Projected mean rice yield (Kg/ha)[a]				Projected rice production (Million tonnes)			
					Medium emission (RCP4.5)		High emission (RCP8.5)		Medium emission (RCP4.5)		High emission (RCP8.5)	
					MC	EC	MC	EC	MC	EC	MC	EC
Northern	10.06	3218	32.38	10.28	2860	2697	2767	2199	29.39 (−9.23)	27.72 (−14.40)	28.43 (−12.19)	22.59 (−30.22)
Central	3.54	1704	6.04	3.49	1451	1211	1433	840	5.06 (−16.17)	4.23 (−30.02)	5.00 (−17.24)	2.93 (−51.47)
Eastern	16.63	2034	33.81	16.59	2062	2053	2067	2082	34.20 (1.15)	34.05 (0.72)	34.28 (1.38)	34.53 (2.11)
Western	0.79	2076	1.64	0.80	2048	2019	2046	1981	1.64 (0.27)	1.62 (−1.12)	1.64 (0.20)	1.59 (−2.99)
Southern	7.42	2896	21.49	7.44	3220	3491	3250	3892	23.94 (11.41)	25.96 (20.79)	24.16 (12.44)	28.94 (34.65)
All India	44.14	2416	106.65	44.32	2345	2289	2332	2174	103.93 (−2.55)	101.45 (−4.87)	103.33 (−3.11)	96.33 (−9.67)

Figures in parentheses indicate % change in rice production over the current production
MC denotes mid-century (2021–2050) and EC denotes end-century (2071–2100)
Current area, yield and production refer to the year 2013–14
[a]The projected mean yield was calculated from the current yield by using the percentage change in mean yield in different regions of India. The future production was worked out by multiplying the projected rice area with a projected mean yield

reduction in yield variability of 2.1%, whereas RCP8.5 scenario predicts an increase in yield variability of 25.9%.

In the western region, as per RCP4.5 and RCP8.5 scenarios, there would be only marginal reduction of rice yield during the periods 2021–2050 and 2071–2100. The changes in mean yield show a variation of 1.9–5.6% during the first period, whereas during the second period, it shows a reduction of 5.8% as per RCP4.5 and an increase of 14.7% as per RCP8.5. The model projects drastic yield variability under both scenarios. The predicted percentage of yield variability as estimated by RCP8.5 scenario is more than that predicted by RCP4.5 scenario for both periods.

In the central region, as per RCP4.5 and RCP8.5 scenarios, there would be a substantial reduction in percentages of rice yield during the periods 2021–2050 and 2071–2100. The reduction in mean yield shows a variation from 14.85 to 15.94% during the first period as per the two scenarios and as per RCP8.5 scenario, it varies from 28.92 to 50.71% during the second period. As expected, yield variability as estimated by RCP8.5 scenario was more than that predicted by RCP4.5 scenario for both periods. The first scenario predicts an increase of 184% yield variability during 2021–2050 while the second scenario predicts an increase of 240% for the same period. For the second period, RCP4.5 predicts an increase of 344% in yield variability, whereas RCP8.5 scenario predicts an increase in yield variability of 667%. The variations are quite large and hence need more careful considerations for planning any future measures.

In the eastern region, as per RCP4.5 and RCP8.5 scenarios, there will be a marginal increase in rice yield during the periods 2021–2050 and 2071–2100. The increase in mean yield varies from 1.40 to 1.63% during the first period as per the two scenarios and it varies from 0.96 to 2.36% during the second period as per the two scenarios. Yield variability as estimated by RCP8.5 scenario is more than that predicted by RCP4.5 scenario for both periods. The first scenario predicts an increase of 35% yield variability during 2021–2050 while the second scenario predicts an increase of 41% for the same period. For the second period, RCP4.5 predicts a higher increase in yield variability of 83%, whereas RCP8.5 scenario predicts an increase in yield variability of 139%. Thus, we could conclude that climate change may not have an adverse effect on the mean rice yield in eastern region but yield variability may be substantial. Hence, measures are to be taken to decrease the yield variability or initiate policy and trade measures to compensate for the yield variations.

In the southern region, as per RCP4.5 and RCP8.5 scenarios, climate change has a positive effect on rice yield. Overall, there will be an increase in rice yield during the periods between 2021–2050 and 2071–2100 under both the scenarios. The percentage increase in mean yield shows a variation from 11.2 to 12.23% during the first period as per the two scenarios, and it varies from 20.57 to 34.40% during the second period as per the two scenarios. Yield variability is expected to reduce in both periods under the two scenarios. The predicted reduction in variability is more with respect to the RCP8.5 scenario. The first scenario predicts a decrease in variability of 9.1% during 2021–2050 while the second scenario predicts a decrease of 20.3% for the same period. For the second period, RCP4.5 predicts a reduction of 18% in yield variability whereas RCP8.5 scenario predicts 101% reduction in variability. Thus, we find that climate change would increase yield accompanied by a substantial reduction in yield variability in the region.

In the southern region, rice production will increase more in the end-century under both medium and high emission scenarios showing the stability of rice production in the region. Being the traditional rice belt of the country, it is crucial to know how best the stability in rice production could be achieved through appropriate measures such as the adoption of different climate resilient practices in rice. Similarly, in the eastern region, rice production is expected to increase marginally and hence efforts should be needed to stabilize the rice production in the region as well. As both southern and eastern regions combined represent 52% of the rice production in the country, it is a good signal that climate change will not have any significant negative impact on rice production. Other regions, accounting for 38% of rice production in the country, will be facing a substantial climate change impact on rice production. It is possible that this impact can be addressed through several adaptation and policy measures so that rice production or supply in the market can be maintained to meet the demand.

Assuming medium emission scenarios will be prevalent in the future, rice production will be approximately 104 million tonnes during the mid-century (2021–2050) and 101 million tonnes during the end-century (2071–2100) indicating that there will be an overall reduction in rice production by 2.5% during the mid-century period and about 5% during end-century period (Table 11.1). The mid-century scenario is not alarming in terms of production. However, the future focus of rice production in the country should be on sustainable practices that will reduce the impact of climate change on the environment. Moreover, efficiency in input usage has to be improved, thus emphasizing not only on climate adaptation but also climate mitigation measures in rice production. This is possible given that there is a potential to improve the efficiency of current technologies used in rice farming systems in the country that can contribute to reduce the greenhouse gas emissions. This requires policy and investment support to reduce the emissions in the mid- and end-century scenarios. Hence, a 5% reduction in rice production in the future can be avoided if proper climate change-related policies are implemented.

Overall, the results indicated a nonlinear effect of both maximum temperature and rainfall in reducing the yield of rice crop in most of the rice growing regions. Further, the variations in yield could increase with an increase in the average maximum temperature.

To summarize, in southern region, climate change will induce an increase in yield accompanied by a substantial reduction in yield variability, whereas in the eastern region, there will be a marginal increase in rice yield accompanied by high increase in yield variability. Though these two regions offer more scope for increasing rice production in the future, eastern region's rice production is subject to the risk of high yield variability which needs to be addressed through technological interventions. In the case of northern, western and central regions, there will be a marginal reduction in rice yield accompanied by medium to high variability in rice yield during the mid- and end-century periods. Among these three regions, central region is more prone to climate change and variability, and hence, special attention is needed for stabilizing rice production as rice area and productivity are already showing declining growth rate in the region. Suitable adaptation strategies have been examined in detail in different regions and those showing promising responses at the farm level have to be promoted. In what follows, possible interventions and approaches are suggested.

11.3 A Way Forward

11.3.1 Future Rice Production Policy

The Population Foundation of India projected that the country's population will be 1546 million, 1695 million and 1824 million by the end of the year 2030, 2040 and 2050, respectively (CRRI 2013). Accordingly, it is estimated that the domestic demand for rice will be 121.2, 129.6 and 137.3 million tonnes during the end of the above periods. In addition to this, India exports about 3.5 million tonnes of basmati and 6.9 million tonnes non-basmati rice every year (CRRI 2013). To meet out the above challenges, the productivity of rice has to be increased, i.e. from current level of 2.4 tonnes per ha to 3.3 tonnes per ha. This should not be difficult to achieve with improvement in the technologies and introduction of new rice varieties although the growth rate in production during the time periods between 2007–08 and 2011–12 was 0.36%, far below the population growth rate of 1.63%. No doubt, the current decelerating trend in production and yield is of concern and hence has to be necessarily reversed to meet the growing demand and also to meet the needs of export markets. Moreover, the profit margin in rice cultivation has been declining and thus making cultivation of this crop unattractive to farmers, especially to the smallholders, who constitute the majority of rice growing farmers in the country. Therefore, a two-pronged strategy of developing new technologies, by expanding research investments to bring efficiency in production, and implementation of favourable government policies will help in increasing rice production and productivity in the country to meet the demand arising in the future. As there will be an only a marginal increase (0.4%) in rice area in the future, any increase in rice production should necessarily come from productivity increases only. Further, adaptation strategies will help improve the rice productivity to a larger extent as indicated in the region-wise chapters of this book.

11.3.2 Upscaling and Mainstreaming the Adaptation Strategies

The evidence of increased climate variability are though obvious, the severity, however, demands more attentions from water and agricultural managers. For new weather patterns repeating for several subsequent seasons, farmers have to adapt strategies to maintain their crop productivity. The reassuring aspect of this argument is that these adaptation options to cope with climate variability would also help to minimize the impacts from other factors like population, urbanization, change in consumer and market demand, poverty, and climate change in the future. Adaptation will benefit from improved forecasting and climate modelling which in turn imply the need for strengthening of data collection initiatives. Agriculture sector needs better tools for prediction and monitoring and introducing sustainable and efficient measures.

Since climate change is a continuous phenomenon and varies between regions, any adaptation and mitigation measures have to be specific in order to minimize the reduction and as well as variability (risk) in rice yield. Adaptations are nothing but adjustments or interventions which are followed to manage the losses or to take advantage of the opportunities offered by climate change (IPCC 2001, 2007). Most of the adaptations that occur at farm-level focus mainly on farming-related interventions or adjustments. Further, they are related to short-term periods and influenced by seasonal climate variations. The earlier chapters have indicated a wide range of adaptation strategies specific to regions. Some of the strategies are direct seeding of rice, modified system of rice intensification (SRI), supplemental irrigation, alternate wetting and drying and improved management and extension practices.

However, access to and adoption of these strategies by farmers varied largely between regions and between states (Palanisami et al. 2014). Hence, region-specific appropriate and affordable adaptation strategies need to be identified for upscaling by the respective state governments since agriculture is a state subject in India. Trade off agricultural produce between the states should be made easier. Transaction cost of the adaptation strategies is a major component in technology adoption. Transaction costs refer to the costs incurred while taking the technology to the fields. Normally, it accounts for 2–5% of the total cost of production of crops and minimizing the transaction cost at farm level will help to speed up the adoption of technologies and this should be well addressed by the extension agencies while promoting various adaptation strategies in different locations. Technology upscaling is important and should be supported by the convergence of several ongoing government programs. Developing public–private partnership in implementation, awareness creation and designing need specific capacity-building training programmes for the extension workers and farmers would improve adoption levels. Climate-smart technologies and a package of practices should be included in all the crop production manuals of the state agriculture departments and universities and made available in local languages.

Further, risk in rice production (in terms of high variability in rice yield) is expected to increase more due to climate change and will influence the rice production in all the regions. Hence, necessary weather-based crop insurance products mostly available for rainfed crops can also be introduced to rice to address the risk in rice production (David et al. 2012).

It is estimated that the adaptation practices namely system of rice intensification, machine transplantation, alternate wetting and drying and direct seeding could reduce the water and labour use by 10–15% and stabilize rice production in the long term. Hence, simply by promotion of these adaptation practices will minimize the income losses to the farmers. Moreover, there is a need for technology upscaling which should be backed up with well-planned capacity-building programmes for the farmers. Further, an assessment of farm-level adaptation strategies over time and space is important in providing information necessary to the implementing agencies including government for formulating the needed policies. The adaptation strategies related to rice production include physical investment to create farm structure as well as management technologies such as improved crop and water management practices, several of which have been already studied by the researchers (Palanisami et al. 2011; Abraha and Savage 2006; Till et al. 2010). Besides, subjecting the farmers to the actual cost of adaptation vis-à-vis to the expected income associated with adaptation measures may motivate them towards better adaptation measures.

11.3.3 Government's Commitment to Address Climate Change

The release of the first National Action Plan on Climate Change (NAPCC) on climate mitigation and adaptation through sustainable development in 2008 evinced India's commitment to adapt to climate change. The plan has identified eight core 'National Missions' to be followed and directed related Ministries of Government to share detailed implementation plans to the Prime Minister's Council on Climate Change. India is thus one of the first countries to come out with NAPCC. However, considering the vast area, the diverse nature of the different states in the country and also the varied impacts of climate across ecosystems and sectors, it is important to device specific strategies within each state, build institutional capacity and develop measures to implement the adaptation plans. Towards this end, the state governments are being sensitized at different levels and expected to device the state climate change adaptation frameworks under the overall supervision of the Ministry of Environment and Forests (MoEF), Government of India. Within the states, the responsibility has been given to the Department of Environment and Forests. One of the primary gaps connected to the planning exercise is the lack of adequate scientific/or technological-based inputs to the state climate and sector adaptation plans. The results on impact due to climate change on the rice production outlined in the book will help to address some of these gaps.

The National Mission for Sustainable Agriculture and the National Water Mission have been identified as the most critical missions in the country. These missions are critical because they hold millions of small-scale farmers in India and their livelihoods, who are highly vulnerable to climate change. The National Mission on Sustainable Agriculture aims to devise strategies to make agriculture more resilient to climate change. The National Water Mission aims to improve water use efficiency by 20%, explore the possibilities of basin-level management strategies in Indian rivers and capacity-building options. Further, these missions have to be supported with inputs from information and biotechnology, as well as credit and insurance support mechanisms. In the process, the respective state governments need support to develop and implement comprehensive adaptation and policy frameworks. Some of them have initiated measures, but need expertise and investments, especially in terms of integrating adaptation technologies with larger development programs and further upscaling.

11.4 Specific Measure Relevant to Rice

11.4.1 Augmenting and Managing Water Resources

The scarcity of water is expected to become an ever-increasing problem in India as a result of climate change impacts. The situation is quite alarming and cautions that India needs to take drastic measures to improve water management, especially groundwater usage (World Bank 2018). The water balance of the regions will change in the future due to the accelerated rate of evaporation from soil and water bodies, change in run-off and soil moisture storage and increased rate of transpiration from plants. Furthermore, it is expected that groundwater depletion will take place due to the combination of shorter duration but more intense rainfall (meaning more run-off and less infiltration) combined with increased evapo-transpiration and increased irrigation (Konikow and Kendy 2005). Lowering of groundwater table and consequently an increase in the energy needed to pump water will make irrigation more expensive particularly for rice in selected regions. Therefore, the strategies and tools must be strengthened to be able to adapt to climate change and to cope with between and within crop season weather variability. In future, a larger amount of water (ranging from 1200 to 2500 mm per hectare) will be needed since India has to produce more rice to meet food grains need for consumption and export. Currently, to produce 1 kg of rice, about 2500–3500 l of water (water productivity) is used in India. Therefore, to meet current challenge of declining water resource base and competition from other sectors, the water productivity of rice needs to be improved by bringing down the water productivity to the level of 2000 L/kg of rice (CRRI 2013; Deelstra et al. 2016; Kakumanu et al. 2016). Our recent research findings show that by using improved crop and water management practices, it is possible to increase the rice productivity and bring down the water use significantly thus improving the water use efficiency (WUE). For e.g., in Krishna, Godavari and Cauvery basins, the average WUE under SRI is 9.1 kg/ha/mm compared to 4.6 kg/ha/mm under conventional practices. Similarly, with alternative wetting and drying (AWD) method, the average WUE is 4.2 kg/ha/mm compared to 2.5 kg/ha/mm under conventional practices (Palanisami et al. 2014; Kakumanu et al. 2016).

In order to achieve the goal of improving the water availability and to meet the demand by 2050, the Ministry of Water Resources, Government of India, is taking several steps. This very optimistic goal is expected to be achieved by several measures, including the implementation of integrated water resource management

(IWRM) plans, enhancing efficiency in water use and demand management (National Water Mission 2008). However, to reach the goal and ensure implementation, substantial investments and cooperation by the governments are necessary.

11.4.2 Managing Soil Health

At the farm level, small-scale farmers are interested in improved soil and land management technologies that can also benefit them with carbon credits. They need technological support to address the challenges in soil management.

The different land-use and soil management options will concomitantly contribute to restoring soil organic matter (SOM) stocks (Lal 2004). Such soil organic matter is also intimately linked to climate effects on the water cycle. Sufficient SOM stocks reduce erosion risks linked to intense precipitation events, help conserve water within the soil profile and help retain nutrients in soils, especially in sandy soils. Thus, building SOM stocks in soils is one of the best possible measures to buffer the effects of climate extremes on crop productivity. In addition, by building SOM stocks in soils under their custody, smallholder farmers contribute to carbon sequestration and storage. In future scenarios, smallholder farmers could get credits (i.e. remuneration) for this important environmental service of CO_2 sequestration and carbon storage in their lands. Increased crop productivity, reduced inter-annual variability and remuneration for carbon storage would go a long way in helping smallholder farmers in rural India.

Therefore, the management of land and soil is crucial for optimizing water use efficiency. In other words, it is at the farm level, that changes have to take place. This might be possible by initiating various measures at the village level, including village knowledge centres, and SHGs to provide necessary help to farmers and women with information and resources to adapt to climate change and training and capacity-building initiatives (Kakumanu et al. 2018). The ongoing information access revolution related to the spread of mobile networks and the rapidly developing situation of mobile Internet access devices could prove to be a new means for state institutions and local agencies for public interaction.

11.4.3 Developing Multiple Abiotic Stress-Tolerant Rice with Climate Resilience

The yield of a crop is influenced by biotic stresses, abiotic stresses and nutritional factors either separately or in combinations. The realization of full yield potential is prevented by abiotic stress primarily due to drought, heat, cold, salinity and submergence. Thus, abiotic stresses always occur in combination. For instance, heat and drought stress or submergence and salinity occur simultaneously. The situation will become further complicated as the climate keep changing in future with rapid fluctuations in the occurrence of such stresses. Therefore, tolerance towards stress is a complex phenomenon and no single approach could provide an immediate solution to the multiple environmental stresses that a plant experiences during its lifecycle. Further, a systematic evaluation of novel germplasm and putative stress-related genes under field conditions has to be undertaken to develop a robust set of new germplasm and better adapted to withstand stresses. Further, identification and pyramiding of multiple genes for stress tolerance in high-yielding genetic backgrounds are feasible and the efforts would be practicable to tailor novel rice genotypes in future (CRRI 2013).

11.4.4 Climate Resilient Rice Production Technologies

The impacts due to climate change can be addressed with the following strategies (i) making adjustments in rice production methods and introduction of new rice strains to withstand high temperature, (ii) developing and popularizing multiple stress-tolerant varieties to withstand and perform better, (iii) adopting environment-friendly and least environment footprint rice-based cropping system, (iv) achieving a sustainable rice production, development of cultivation practices to maintain soil health and conserve natural resource base has to be promoted, (v) accelerating adaptation to climate change and better management of agricultural risks towards resilient agriculture and (vi) promoting partnership with rural communities and other stakeholders in a climate-smart model for agricultural development along with innovative agricultural risk management measures (CRRI 2013).

The National Rice Research Institute (NRRI) in its Vision 2050 clearly states that entrepreneurship development is important. By providing specialized training to educated rural youth, one can encourage and make them successful entrepreneurs. The trained youths, in the long run, will act as change agents to disseminate and promote the adoption of rice-related interventions including spread of high-yielding varieties (HYV) and seed production, production and marketing of rice, farm mechanization, value-added rice products, etc. (CRRI 2013). Further, strengthening gender participation needs to be ensured. With much emphasis on women empowerment and sensitization on gender equality and equity, rice researchers have to develop projects for generation of technology that would suit the needs, aspiration and problems of women in agriculture. Special attention on the drudgery of women in rice farming needs to be made such that the occupational hazards while transplanting, weeding and post-harvest can be mitigated and ultimately the efficiency of women in rice production can be increased.

11.5 Summary

The preceding discussions on climate change clearly indicate that such changes would lead to a negative impact on crop yields in most of the regions in India. One way of addressing these impacts will be through the adoption of feasible strategies across the rice value chain that will help in sustaining rice production. Since the current technology adoption level is comparatively low (ranging from 8 to 12%) more efforts are needed to promote the strategies to be adopted by the farmers (Palanisami et al. 2013, 2015). Therefore, future research and policy interventions have to focus on studying the economics of different adaptation strategies already being followed by the farmers as well as the new strategies that are being field tested in different locations by different agencies. Moreover, studies on farmers' responses to these strategies are also warranted as the most efficient strategies in terms of profit are not followed by farmers owing to operating constraints in adopting them. Provisions of assured and timely credit, marketing facilities with remunerative prices and weather-based crop insurance products will help improve the technology adoption. Public and private sector partnership in promoting the interventions need to be strengthened. In this context, the following recommendations are important:

- Capacity building to farmers, through awareness programs, to educate them on possible impacts on climate change, improving adaptation strategies and the effective use of support systems. The development and implementation of the adaptation measures need to be done through 'learning-by-doing'.
- Build creative partnerships to facilitate adaptation through stakeholder participation; work closely with local institutions; demonstrate multiple adaptation strategies; and carry out a cost-benefit analysis of adaptation strategies. A careful economic analysis of the instruments will facilitate adaptation, strengthening farmer producer groups/organizations to support farm mechanization, trade and better price to rice farmers.
- Crop diversification for effective use of land and water, which in turn will ensure better income to rice farmers.
- Existing weather-based insurance products to be fine-tuned according to the regions and crops.
- Use of ICT-based tools, mobile phones and rice apps to improve farmer connectivity and rice productivity. Several states have already developed user-friendly software/portals to facilitate the use of online extension services to reach larger farmer numbers.
- The availability of relevant data/information must be ensured and such information is very crucial for making climate projections and identifying vulnerable groups and hot spots.
- Information sharing, including the early warning system, between sectors is essential for developing effective and efficient climate change adaptation strategies, both in the short and long term.
- State level initiatives like 'Climate Water Forum' as was done in Andhra Pradesh state, India will help to address key emerging issues and make them more sustained.

- Policy and trade measures to make it easier for farmers to sell their produce. This includes assuring acceptable minimum support price and facilities to store and market the harvested products to minimize post-harvest losses.

References

Abraha MG, Savage MJ (2006) Potential impacts of climate change on the grain yield of maize for the midlands of KwaZulu-Natal, South Africa. Agric Ecosyst Environ 115:150–160

African Devlopment Bank et al. (2009). Poverty and climate change: reducing the vulnerability of the poor through adaptation. Available at http://documents.worldbank.org/curated/en/534871468155709473/pdf/521760WP0pover1e0Box35554B01PUBLIC1.pdf

Adhikari BM, Bag K, Bhowmick MK, Kundu C Status paper on rice in West Bengal. Retrieved from http://www.rkmp.co.in

Aggarwal PK (2008) Global climate change and Indian agriculture: Impacts, adaptation and mitigation. Indian J Agric Sci 78(10):911–919

Aggarwal PK, Hebbar KB, Venugopalan MV, Rani S, Bala A, Biswal A, Wani SP (2008) Quantification of yield gaps in rain-fed rice, wheat, cotton and mustard in India. Global theme on agro-ecosystems report no. 43. International Crops Research Institute for the Semi-Arid Tropics. Patancheru, Andhra Pradesh, India

Aggarwal PK, Singh AK, Samra JS, Singh G, Gogoi AK, Rao GGSN, Ramakrishna YS (2009) Introduction. In: Aggarwal PK (ed) Global climate change and Indian agriculture. ICAR, New Delhi, pp 1–5

Aggarwal PK, Kropff MJ, Cassman KG, Berge HFM (1997) Simulating genotypic strategies for increasing rice yield potential in irrigated, tropical environments. Field Crops Res 51:5–17

Aggarwal PK, Mall RK (2002) Climate change and rice yields in diverse agro-environments of India: effect of uncertainties in scenarios and crop models on impact assessment. Clim Change 52(3):331–343

Aggarwal PK, Pathak H, Kumar N (2009) Global climate change and indian agriculture: a review of adaptation strategies. Trust for Advancement of Agricultural Sciences, New Delhi

Agricultural Statistics (2014) Agricultural statistics at a glance. Directorate of economics and statistics. Government of India

Agricultural Statistics (2016) Agricultural statistics at a glance. Directorate of economics and statistics. Government of India

Akaike H (1973) Information theory and an extension of the maximum likelihood principle. In: Petrov BN, Csaki F (eds) Proceedings of the 2nd international symposium on information theory. Akademiai Kiado, Budapest, pp 267–281

Anderson J, Hazell P (1989) Variability in grain yields: implications for agricultural research and policy in developing countris. Intrnational Food Policy Research Institute. The Johns Hopkins University Press. Baltimore and London

Annamalai H, Hafner J, Sooraj KP, Pillai P (2013) Global warming shifts the monsoon circulation, drying South Asia. J Clim 26(9):2701–2718

K. Palanisami et al., *Climate Change and Future Rice Production in India*, India Studies in Business and Economics,
https://doi.org/10.1007/978-981-13-8363-2

Asche F, Tveterås R (1999) Modeling production risk with a two-step procedure. J Agric Resour Econ 24:424–439

Aye GC, Ater PI (2012) Impact of climate change on grain yield and variability in Nigeria: a stochastic production model approach. Mediterr J Soc Sci 3(16):142–150

Bal PK, Ramachandran A, Palanivelu K, Thirumurugan P, Geetha R, Bhaskaran B (2016) Climate change projections over India by a downscaling approach using PRECIS. Asia Pacific J Atmos Sci 52(4):353–369

Baltagi BH (2005) Econometric analysis of panel data, 3rd edn. Wiley, New York

Barnwal P, Kotani K (2010) Impact of variation in climate factors on crop yield: a case of rice crop in Andhra Pradesh, India. Working paper. IUJ Research Institute, International University of Japan, Minami-Uonuma

Barton DN, Kakumanu KR, Palanisami K, Tirupataiah K (2012) Analysis of economic incentives to manage risk at farm level in the context of climate change. In: Nagothu US, Gosain AK, Palanisami K (eds) Water and climate change: an integrated approach to address adaptation challenges. Macmillan Publishers India Ltd., New Delhi, pp 143–168. ISBN 978-935-059-059-1

Basavaraja H et al (2008) Technological change in paddy production: a comparative analysis of traditional and sri methods of cultivation. Indian J Agric Econ 63(4)

Bindi M, Fibbi L, Bernardo Gozzini, Serena Orlandini, Miglietta F (1996) Modelling the impact of future climate scenarios on yield and yield variability of grapevine. Clim Res 7:213–224. https://doi.org/10.3354/cr007213

Boubacar I (2010) The effects of drought on crop yields and yield variability in Sahel. Int J Econ Finance 4(12)

Buchinsky M (1998) Recent advances in quantile regression models: a practical guideline for empirical research. J Human Resour 33(1):88–126. Retrieved from https://EconPapers.repec.org/RePEc:uwp:jhriss:v:33:y:1998:i:1:p:88-126

Bouman B, Barker R, Humphreys E, Tuong TP, Atlin G, Bennett J, Dawe D, Dittert K, Dobermann A, Facon T, Fujimoto N, Gupta R, Haefele S, Hosen Y, Ismail A, Johnson D, Johnson S, Khan S, Shan L, Masih I, Matsuno Y, Pandey S, Peng T, Muthukumarisami T, Wassman R (2007) Rice: feeding the billions. In: Molden D (ed) Water for food, water for life: a comprehensive assessment of water management in agriculture. Earthscan/IWMI, London, UK/Colombo, Sri Lanka, pp 515–549

Cabas-Monje JW, Weersink AO, Edward O (2010) Crop yield response to economic sit and climatic variables. Clim Change 101:599–616. https://doi.org/10.1007/s10584-009-9754-4

Cambell B, Mannm W, Melendez-Ortiz R, Streck C, Tennigkeit T (2011) Addressing agriculture in climate change negotiations: a scoping report. Retrieved from https://ccafs.cgiar.org/publications/addressing-agriculture-climate-change-negotiations-scoping-report#.XATlSPZuLIU. Accessed on 3 Dec 2018

Central Rice Research Institute (CRRI) (2013) Vision 2050. Indian Council of Agricultural Research, Cuttack, India

Chapagain T, Riseman A, Yamaji E (2011) Achieving more with less water: Alternate wet and dry irrigation (AWDI) as an alternative to the conventional water management practices in rice farming. J Agric Sci 3(3):3–13

Chaturvedi RK, Joshi J, Jayaraman M, Govindasamy B (2012) Multi-model climate change projections for India under representative concentration pathways. Curr Sci 103(7):791–802

Chen C-C, Chang CC (2005) The impact of weather on crop yield distribution in taiwan: some new evidence from panel data models and implications for crop insurance. Agric Econ 33 (s3):503–511

Chen C-C, McCarl BA, Schimmelpfennig D (2004) Yield variability as influenced by climate: a statistical investigation. Clim Change 66(2):239–261

Chen S, Cai SG, Chen X, Zhang GP (2009) Genotypic differences in growth and physiological responses of transplanting and direct seeding cultivation in rice. Rice Sci 16(2):143–150

Cheralu C (2011) Status paper on rice in Andhra Pradesh. Rice Knowl Manag Portal 1–36. Retrieved from http://www.rkmp.co.in/sites/default/files/ris/rice-state-wise/RiceState Wise Andhra Pradesh_0.pdf

Cherchi A, Annamalai H, Masina S, Navarra A (2014) South Asian summer monsoon and the eastern Mediterranean climate: the monsoon-desert mechanism in CMIP5 simulations. J Clim 27(18):6877–6903

Choi I (2001) Unit root tests for panel data. J Int Money Finance 20(2):249–272

Cline WR (2007) Global warming and agriculture: impact estimates by country. Peter Institute for international economics, Columbia University Press, 250 pp

Das SR (undated) Status paper on rice in Orissa. Rice Knowledge Management Portal, Directorate of Rice Research, Hyderabad. Retrieved from http://www.rkmp.co.in/sites/default/files/ris/rice-statewise/Status%20Paper%20on%20Rice%20in%20Orissa.pdf

De-Graft AH, Kweku KC (2012) The effects of climatic variables and crop area on maize yield and variability in Ghana. Russ J Agric Soc-Econ Sci 10:10–13. https://doi.org/10.18551/rjoas.2012-10.02

Deelstra J, Kakumanu KR, Kallam SR, Nagothu US, Geethalakshmi V, Lakhsman A, Arasu MS (2016). Water productivity under different rice growing practices: results from farmer led field demonstrations in India. In: Nagothu US (eds) Climate change and agricultural development: improving resilience through climate smart agriculture, agroecology and conservation. Earthscan Food and Agriculture, Routledge, UK. ISBN 978-1-138-92227-3

Deressa T, Hassan R (2009) Economic impact of climate change on crop production in ethiopia: evidence from cross-section measures. J Afr Econ 18:529–554. https://doi.org/10.1093/jae/ejp002

Dey MM, Upadhyaya HK (1996) Yield loss due to drought, cold, and submergence in Asia. In: Evenson RE, Herdt RW, Hossain M (eds) Rice research in Asia: progress and priorities. CAB International/IRRI, Wallingford, UK, pp 35–58

Dwivedi JL (2011). Status paper on rice in Uttar Pradesh. Rice Knowledge Management Portal. Retrieved from http://www.rkmp.co.in

El-Ashry M (2009) Adaptation to climate change: building resilience and reducing vulnerability. Recommendations from the 2009 Brooking Blum Roundtable, United Nations Foundations

Farooq M, Kadambot HMS, Rehman H, Aziz T, Lee D, Wahid A (2011) Rice direct seeding: experiences, challenges and opportunities. Soil Tillage Res 111(2011):87–98

Food and Agriculture Organization of the United Nations (FAO) (1992) Agrostat. FAO Publication, Rome, Italy

Food and Agriculture Organization of the United Nations (FAO) (1996) Agro−ecological zoning: guidelines FAO Soils Bulletin 73, Rome, Italy

Food and Agricultural Organization of the United Nations (FAO) (2000) Two essays on climate change and agriculture: a developing country perspective. FAO Economic and Social Development Paper, p 145

Food and Agriculture Organization of the United Nations (FAO) (2009) OECD-FAO Agricultural outlook 2009–2018

Garrett HE, Woodworth RS (1971) Statistics in psychology and education. Vakils, Feffer and Simons, Bombay

Geethalakshmi V, Lakshmanan A, Rajalakshmi D, Jagannathan R, Sridhar G, Ramaraj AP, Bhuvaneswari K, Gurusamy L, Anbhazhagan R (2011) Climate change impact assessment and adaptation strategies to sustain rice production in Cauvery Basin of Tamil Nadu. Curr Sci 101:342–347

Government of India (2004) National action plan on climate change in India. Retrieved from www.moef.nic.in/downloads/home/pg01-52.pdf

Government of India (2012) Twelfth five year (2012–2017) economic sectors, vol II. Government of India, p 3

Granger CWJ, Newbold P (1974) Spurious regressions in econometrics. J Econ 2(2):205–210

Gujja B, Thiyagarajan TM (2009) New hope for indian food security? The system of rice intensification. Gate Keeper 143, International Institute for Environment and Development, London

Gujarati DN (2004) Basic econometrics. McGraw Hill, New York

Gupta S, Sen P, Srinivasan S (2012) Impact of climate change on Indian economy: evidence from food grain yields. Centre for Development Economics Working Paper 218, Delhi

Haris AA, Biswas S, Chhabra V (2010) Climate change impacts on productivity ofrice (*OryzaSativa*) in Bihar. Indian J Agron 55:295–298

Harvey AC (1976) Estimating regression models with multiplicative heteroscedasticity. Econometrica 44:461–465

Hasanthika WKAMA, Edirisinghe JC, Rajapakshe RDDP (2013) Climate variability, risk and paddy production. J Environ Prof Sri Lanka 2(2):57–65

Hoff H (2004) Climate change adaptation in India. Research report. Postdam Institute for Climate Impact Research, Germany

Hou AH, Chen GX, Wang ZP, Cleemput OV, Patrick WH (2000) Methane and nitrous oxide emissions from a rice field in relation to soil redox and microbiological processes. Soil Sci Soc Am J 64(6):2180–2186

Holst R, Yu X, Grun C (2010) Climate change, risk and grain production in China. In: 2010 AAEA, CAES, WAEA joint annual meeting. Agriculture and Applied Economics Association, Colorado, p 29

Hundal SS, Kaur P (2007) Climatic variability and its impact on cereal productivity in Indian Punjab. Curr Sci 92:506–512

Hurlin C, Mignon V (2006) Second generation panel unit root tests. Retrieved from http://halshs.archives-ouvertes.fr/docs/00/15/98/42/PDF/UnitRoot_Ev5.pdf

Im KS, Pesaran MH, Shin Y (2003) Testing for unit roots in heterogeneous panels. J Econ 115:53–74

Indiastat (2013) Stat-wise production of rice in India. Retrieved from www.Indiastat.com

IPCC (2001) Climate change 2001: impacts, vulnerability and adaptation. In: Contribution of working group III to the third assessment report on the Intergovernmental Panel on Climate Change. Cambridge University Press, Cambridge

IPCC (2007) Climate Change 2007: the physical science basis: summary for policymakers working group I to the Intergovernmental Panel on Climate Change fourth assessment report. Intergovernmental Panel on Climate Change, Geneva

IPCC (2007) Climate Change 2007: Synthesis Report. Contribution of Working Groups I, II and III to the Fourth Assessment Report of the Intergovernmental Panel on Climate Change [Core Writing Team, Pachauri, R.K and Reisinger, A. (eds.)]. IPCC, Geneva, Switzerland, 104 pp

IPCC (2014) Climate change: impacts, adaptation and vulnerability. Cambridge University Press, Cambridge

IPCC (2018) Summary for policymakers. In: Masson-Delmotte V, Zhai P, Pörtner HO, Roberts D, Skea J, Shukla PR, Pirani A, Moufouma-Okia W, Péan C, Pidcock R, Connors S, Matthews JBR, Chen Y, Zhou X, Gomis MI, Lonnoy E, Maycock T, Tignor M, Waterfield T (eds) Global warming of 1.5 °C. An IPCC special report on the impacts of global warming of 1.5 °C above pre-industrial levels and related global greenhouse gas emission pathways, in the context of strengthening the global response to the threat of climate change, sustainable development, and efforts to eradicate poverty. World Meteorological Organization, Geneva, Switzerland, 32 pp

Isik M, Devadoss S (2006) An analysis of the impact of climate change on crop yields and yield variability. Appl Econ 38:835–844

Isik M, Khanna M (2003) Stochastic technology, risk preferences and adoption of site−specific technologies. Am J Agric Econ 85:305–317

Joshi E, Dinesh K, Lal B, Nepalia V, Gautam P, Vyas AK (2013) Management of direct seeded rice for enhanced resource—use efficiency. Plant Knowl J 2(3):119–134

Jung H-S, Choi Y, Oh J-H, Lim G-H (2002) Recent trends in temperature and precipitation over South Korea. Int J Climatol 22:1327–1337

Judge GG, Hill RC, Griffiths WE, Lutkepohl H, Lee TC (1988) Introduction to the theory and practise of econometrics, 2nd edn. Wiley, New York

Just RE, Pope RD (1978) Stochastic specification of production functions and economic implications. J Econometrics 7:67–86

Kakumanu KR, Kotapati GR, Nagothu US, Palanisami K, Kallam SR (2018) Adaptation to climate change and variability: a case of direct seeded rice in Andhra Pradesh. J Water Clim Change, India. https://doi.org/10.2166/wcc.2018.141

Kakumanu KR, Kaluvai YR, Nagothu US, Lati NR, Kotapati GR, Karanam S (2018) Building farm-level capacities in irrigation water management to adapt to climate change. Irrig Drainage 67:43–54. https://doi.org/10.1002/ird.2143

Kakumanu KR, Tesfai M, Borrell A, Nagothu US, Suresh Reddy K, Gurava Reddy K (2016) Climate smart rice production systems: studying the potential of alternate wetting and drying irrigation. In: Nagothu US (eds) Climate change and agricultural development: Improving resilience through climate smart agriculture, agroecology and conservation. Earthscan Food and Agriculture, Routledge, UK. ISBN 978-1-138-92227-3

Kelkar U, Bhadwal S (2007) South Asian regional study on climate change impacts and adaptation: implications for human development. In: Occasional paper. Human Development Report Office, United Nations Development Programme, New Delhi, India

Kendall MG, Babington Smith B (1939) The problem of *m* rankings. Ann Math Stat 10(3): 275–287

Kim M-K, Pang A (2009) Climate change impact on rice yield and production risk. J Rural Dev Nongchon-Gyeongje 2:1–13. Korea Rural Economic Institute

Konikow LF, Kendy E (2005) Groundwater depletion: a global problem. Hydrogeol J 13:317–320

Komiya S, Shoji Y, Noborio K, Yazaki T, Toojinda T, Siangliw M (2010) Climatic impacts on greenhouse gas emissions in rice paddy fields. Paper presented at 19th world congress of soil science, soil solutions for a changing world, Brisbane, Australia, 1–6 Aug

Koenker R, Bassett G (1978) Regression quantiles. Econometrica 46(1):33–50

Kulkarni RS, Raghuprasad KP, Mallika M (2013) Achievements, challenges, technology dissemination of rice in India. Report Innovation in rice Production, p 215

Kumar K, Parikh J (1998a) Climate change impacts on Indian agriculture: results from a crop modelling approach. In: Dinar A, Mendelsohn R, Evenson R, Parikh J, Sanghi A, Kumar K, McKinsey J, Lonergan S (eds) Measuring the impact of climate change on Indian agriculture. World Bank Technical Paper No. 402, Washington, DC

Kumar K, Parikh J (1998b) Climate change impacts on Indian agriculture: The Ricardian approach. In: Dinar A, Mendelsohn R, Evenson R, Parikh J, Sanghi A, Kumar K, McKinsey J, Lonergan S (eds) Measuring the impact of climate change on Indian agriculture. World Bank Technical Paper No. 402, Washington, DC

Kumar P, Joshi PK, Birthal PS (2009) Demand projections for food grains in India. Agric Econ Res Rev 22(2):237–243

Kumar P, Shinoj P, Raju SS, Anjani K, Msangi S (2010) Agric Econ Res Rev 23:9

Kumbhakar SC (1997) Efficiency estimation with heteroscedasticity in a panel data model. Appl Econ 29:379–386

Lakshmanan A, Geethalakshmi V, Nagothu US (2009) Azolla and cyanobacterial systems in supplementing nitrogen to rice besides minimizing methane flux. Technical brief 2, CLIMARICE: climate change and persistent droughts: impact, vulnerability and adaptation in rice growing sub-divisions in India. Tamilnadu Agricultural University, Coimbatore

Lal R (2004) Soil carbon sequestration impacts on global climate change and food security. Science 304:1623–1627

Li Y, Simunek J, Jing L, Zhang Z, Ni L (2014) Evaluation of water movement and water losses in a direct-seeded-rice field experiment using hydrus-1D. Agric Water Manage 142:38–46

Levin A, Lin CF, Chu CS (2002) Unit root tests in panel data: asymptotic and finite-sample properties. J Econometrics 108:1–24

Maddala G, Wu S (1999) A Comparative study of unit root tests and a new simple test. Oxford Bull Econ Stat 61:631–652

Mahajan G, Chauhan BS, Gill MS (2013) Dry-seeded rice culture in Punjab state of India: lessons learned from farmers. Field Crops Res 144:89–99

Mall RK, Singh R, Gupta A, Srinivasan G, Rathore LS (2006) Impact of climate change on Indian agriculture: a review. Clim Change 78(2–4):445–478

McCarl BA, Villavicencio X, Wu M (2008) Climate change and future analysis: is stationarity dying? Am J Agric Econ 90(5):1241–1247

Mearns LO, Rosenzweig C, Goldberg R (1997) Mean and variance change in climate scenarios: methods, agricultural applications, and measures of uncertainty. Clim Change 35:367–396

Mehala V et al (2016) Impact of direct seeded rice on economics of paddy crop in Haryana. Int J Agric Sci 8(62):3525–3528. ISSN 0975-3710, E-ISSN 0975-9107

Mendelsohn R, Dinar A, Dalfelt A (2000) Climate change impacts on African agriculture. School of Forestry and Environmental Studies, Yale University, Mimeo

Mendelsohn R, Nordhaus WD, Shaw D (1996) The impact of global warming on agriculture: reply. Am Econ Rev 86:1312–1315

Mendelsohn R, Nordhaus WD, Shaw D (1994) The impact of global warming on agriculture: A Ricardian analysis. Am Econ Rev 84:753–771

Moss RH, Edmonds JA, Hibbard KA, Manning MR, Rose SK, Van Vuuren DP, Carter TR, Emori S, Kainuma M, Kram T, Meehl GA, Mitchell JFB, Nakicenovic N, Riahi K, Smith SJ, Stouffer RJ, Thomson AM, Weyant JP, Wilbanks TJ (2010) The next generation of scenarios for climate change research and assessment. Nature 463(7282):747–756

Nandhini US, Alagumani T, Shibi S (2006) Economic analysis of agriculture in southern parts of coastal India. Agricultura Tropica et Subtropica 39(4):279–284

Oliver MMH, Talukder MSU, Ahmed M (2008) Alternate wetting and drying irrigation for rice cultivation. J Bangladesh Agric Univ 6(2):409–414. ISSN 1810-3030

Paddy cultivation guide: different methods of rice cultivation in India. Retrieved from https://www.farmingindia.in/paddy-cultivation/

Palanisami K, Paramasivam P, Ranganathan CR, Aggarwal PK, Senthilnathan S (2009) Quantifying vulnerability and impact of climate change on production of major crops in Tamil Nadu, India. In: Taniguchi M, Burnett WC, Fukushima Y, Haigh M, Umezawa Y (eds) Headwaters to the Ocean—hydrological change and watershed management. Taylor and Francis Group, London, pp 509–551

Palanisami, K, Suresh Kumar D, Malik, RPS, Raman S, Kar G, Kadiri M (2015) Managing water management research: analysis of four decades of research and outreach programmes in India. Econ Polit Wkly l(26&27)

Palanisami K, Ranganathan CR, Kakumanu KR, Nagothu US (2011) A hybrid model to quantify the impact of climate change on agriculture in Godavari river basin. Energy Environ Res 1 (1):32–52

Palanisami K, Ranganathan CR, Nagothu US, Kakumanu Krishna Reddy (2014) Climate change and agriculture in India. Routledge Taylor and Francis Group, New Delhi. ISBN 978-0-415-73599-5

Palanisami K, Krishna Reddy K, Ranganathan CR, Nagothu US (2015) Farm-level cost of adaptation and expected cost of uncertainty associated with climate change impacts in Major River basins in India. Int J Clim Change Strat Manage 7(1):76–96

Pandey MP, Verulkar SB, Sarawgi AK. Status paper on Rice in Chattisgarghl. Retrieved from http://www.rkmp.co.in

Pathak AR, Dr Mehta AM, Dr. Vashi RD. Retrieved from http://www.rkmp.co.in

Pathak H, Ladha JK, Aggarwal PK, Peng S, Das S, Singh Y, Singh B, Kamra SK, Mishra B, Sastri ASRAS, Aggarwal HP, Das DK, Gupta RK (2003) Climatic potential and on-farm yield trends of rice and wheat in the Indo-Gangetic Plains. Field Crops Res 80(3):223–234

Pathak H, Sankhyan S, Dubey DS, Bhatia A, Jain N (2013) Dry direct-seeding of rice for mitigating greenhouse gas emission: field experimentation and simulation. Paddy Water Environ 11:593–601

Pattanayak A, Kumar KSK (2013) Weather sensitivity of rice yield: evidence from India. Working papers 2013-081. Madras School of Economics, Chennai, India

Pepsico International (2009). Direct seeding of paddy—the work of Pepsico reported in India water portal. Retrieved from http://www.indiawaterportal.org/post/6754. Accessed on 11 Aug 2013

Prakash SS, Patil CR, Amrutha TJ, Siddayya S (2013) Input use and production pattern of paddy cultivation under leased-in land in Tungabhadra project area. Karnataka J Agric Sci 26(2): 224–228

Pattanayak A, Kumar KSK (2013) Weather sensitivity of rice yield: evidence from India. Working papers 2013-081. Madras School of Economics, Chennai. India. Retrieved from https://ideas.repec.org/p/mad/wpaper/2013-081.html

Rajanna MP (2010) Status paper on rice in Karnataka. Directorate of rice research. Rice Knowledge Management Portal (RKMP), Hyderabad, India. Retrieved from http://rkmp.co.in/sites/default/files/ris/rice-statewise/Status%20Paper%20on%20Rice%20in%20Karnataka.pdf

Ranganathan C (2009) Quantifying the impact of climatic change on yields and yield variability of major crops and optimal land allocation for maximizing food production in different agro-climatic zones of Tamil Nadu, India: An econometric approach. Working paper, Research Institute for Humanity and Nature, Kyoto, Japan

Rang A, Mangat GS, Kaur R. Status paper on rice in Punjab. Retrieved from http://www.rkmp.co.in

Rao SK. Status paper on rice in Madhya Pradesh. Retrieved from http://www.rkmp.co.in

Reddy TY, Reddi GH (1995) Principles of agronomy. Kalyani publishers, India

Reiner W, Aulakh MS (2000) The role of rice plants in regulating mechanisms of methane emissions. Biol Fertil Soils 31:20–29

Saha A, Havenner A, Talpaz H (1997) Stochastic production function estimation: small sample properties of ML versus FGLS. Appl Econ 29:459–469

Sahana S et al (2017) Adoption and attitude of the farmers on mechanization of paddy in command areas of Karnataka. Int J Agric Sci 9(29):4382–4384. ISSN 0975-3710, E-ISSN 0975-9107

Sarker MAR, Alam K, Gow J (2012) Exploring the relationship between climate change and rice yield in Bangladesh: an analysis of time series data. Agric Syst Elsevier 112:11–16

Saseendran SA, Singh KK, Rathore LS, Singh SV, Sinha SK (2000) Effects of climate change on rice production in the tropical humid climate of Kerala, India. Clim Change 44:495–514

Schlenker W, Roberts MJ (2008) Estimating the impact of climate change on crop yields: the importance of nonlinear temperature effects. In: NBER working paper 13799

Seo, SN, Mendelsohn R (2008) A structural Ricardian analysis of climate change impacts and adaptations in African agriculture. In: Policy research working paper No. 4603. World Bank, Washington, DC. © World Bank. https://openknowledge.worldbank.org/handle/10986/6770. License: CC BY 3.0 IGO

Shanmugam TR, Mahendran K, UshaNandhini S, Shibi S (2007) Impact assessment of agricultural research and technology in addressing rice productivity constraints in Tamil Nadu. Project report submitted to the Department of Agricultural Economics, Tamil Nadu Agricultural University, Coimbatore 3, India

Shetty PK, Hegde MR, Mahadevappa M (eds) Innovations in rice production. National Institute of Advanced Studies, Bangalore, pp 1–243

Sperber KR, Annamalai H, Kang I-S, Kitoh A, Moise A, Turner A, Wang B, Zhou T (2013) The Asian summer monsoon: an intercomparison of CMIP5 versus CMIP3 simulations of the late 20th century. Clim Dyn 41(9–10):2711–2744

Suresh Kumar D, Palanisami K (2010) Impact of drip irrigation on farming system: evidence from southern India. Agric Econ Res Rev 23:265–272

Thakur J (1994) Rice production constraints in Bihar (India). Paper presented at the workshop. In: Thaware BL, Kunkerkar RL, Shivade HA (eds) Status paper on rice in Maharashtra. Retrieved from http://www.rkmp.co.in

Thiyagarajan K, Kalaiyarasi R. Status paper on rice in Tamil Nadu. Retrieved from http://www.rkmp.co.in

Technical Program (2011) Technical report of agricultural economics. Regional Agricultural Research Station, Lam, Guntur, Andhra Pradesh. Unpublished manuscript

Till B, Artner A, Siebert R, Sieber S (2010). Micro-level practices to adapt to climate change for African small-scale farmers. IFPRI Discussion Paper 00953

UNESCO-WWAP (2009) Climate change and water—an overview from the world water development report 3: Water in the changing world. A world water assessment programme special report, Perguia, Italy

Wang JX, Mendelsohn R, Dinar A, Huang JK, Rozelle S, Zhang LJ (2009) The impact of climate change on China's agriculture. Agric Econ 40(3):323–337

World Bank (2018) India—Atal Bhujal Yojana (ABHY)—National groundwater management improvement program (English). Washington, DC

Zainal Z, Shamsudin MN, Mohamed ZA, Adam SU (2012) Economic impact of climate change on the Malaysian palm oil production. Trends Appl Sci Res 7(10):872–880. https://doi.org/10.3923/366tasr.2012.872.880

Printed by Printforce, the Netherlands